Christian Remling
Spectral Theory of Canonical Systems

De Gruyter Studies in Mathematics

Volume 70

Christian Remling

Spectral Theory of Canonical Systems

DE GRUYTER

Mathematics Subject Classification 2010
Primary: 34L05, 34L40, 46E22; Secondary: 34A55, 47A06

Author
Prof. Dr Christian Remling
University of Oklahoma
Department of Mathematics
Norman 73019-3103
USA
christian.remling@ou.edu

ISBN 978-3-11-056202-6
e-ISBN (PDF) 978-3-11-056323-8
e-ISBN (EPUB) 978-3-11-056228-6
ISSN 0179-0986

Library of Congress Control Number: 2018950596

Bibliographic information published by the Deutsche Nationalbibliothek
The Deutsche Nationalbibliothek lists this publication in the Deutsche Nationalbibliografie; detailed bibliographic data are available on the Internet at http://dnb.dnb.de.

Typesetting: VTeX UAB, Lithuania
Printing and binding: CPI books GmbH, Leck

www.degruyter.com

Contents

Preface

> Facturusne operae pretium sim, si a primordio urbis res populi Romani perscripserim, nec satis scio, nec, si sciam, dicere ausim, quippe qui cum veterem tum vulgatam esse rem videam, dum novi semper scriptores aut in rebus certius aliquid adlaturos se aut scribendi arte rudem vetustatem superaturos credunt. Utcumque erit, iuvabit tamen rerum gestarum memoriae principis terrarum populi pro virili parte et ipsum consuluisse, et si in tanta scriptorum turba mea fama in obscuro sit, nobilitate ac magnitudine eorum me, qui nomini officient meo, consoler.
>
> T. Livius, *Ab urbe condita, praefatio*

Nor am I certain that the effort was worthwhile. The Roman historian Livy wrote these words around 30 BC, and he could not possibly have imagined today's publishing landscape. Scientific books and articles now appear at a rate that would have been completely impossible to grasp for anyone living in the ancient world. It suffices to subscribe to the e-mail service of the electronic preprint repository arxiv.org to get a faint impression of this: the friendly server will now send you an announcement of some five or so new preprints in your own narrow area on a daily basis that, apparently, one has to scan just to stay up to date in a specialized field.

Obviously, the situation has gotten out of hand,[1] and one should have very good reasons indeed before contributing to the problem. While I dare not hope that mine might be judged sufficient by any reasonable standards, I can, by way of apology, offer a few words on how I came to write this book.

Some twenty years ago, I met Seppo Hassi and Henk de Snoo, both experts on canonical systems, at conferences. Henk subsequently invited me for several visits, with the idea to explore topics for a possible collaboration. I'm afraid I wasn't of much use (we did write one joint paper, with Seppo), but I certainly enjoyed those visits and Henk's fantastic hospitality at his home near Groningen, Holland. I was just a few years past my PhD at the time, and I had started working on other differential and difference operators (Schrödinger, Jacobi), but had not heard of canonical systems before; I really didn't know much about anything. When Henk mentioned some of the fundamental results on canonical systems, I was immediately struck by their elegance and power and wanted to know where I could learn more, and Henk's answer to this would invariably be "ah, well, that's due to de Branges." So on to de Branges's papers [13–16] and especially his book [17], and this slowed me down quite efficiently. While it quickly becomes clear that Louis de Branges single-handedly built an amazingly complete and deep theory (and I do wish to go on record as saying that I consider this

1 It could be argued that one can never have too much of a good thing, but this misses the point. Even if one assumed (which I certainly don't), for the sake of the argument, that all published research is of acceptable quality, then it still does not feel right to have so much research published and so little of it read by anyone.

https://doi.org/10.1515/9783110563238-201

an outstanding mathematical achievement), it also cannot be denied that he makes equally gigantic demands on the reader.

I did pick up a few things and also tried to use them later, but I did not really gain a systematic understanding of canonical systems. Since that time, I have always thought that it is a pity that there is no gentler way into this fascinating area that is so central to the spectral theory of differential and difference operators, and the idea of a book on the subject has always been at least in the back of my mind.

So here I am, attempting, *pro virili parte*, to write such a book. Ideally, it should appeal to two completely different types of readers: experts from related areas such as spectral theory who are not yet very familiar with canonical systems, and graduate students and beginning researchers who are just starting out. Experts on canonical systems will of course already be familiar with most of what I do here, but even they might find the very detailed proofs of the fundamental results useful that I give, and the book does contain some original material, especially in the later chapters. But it would really be more honest to say that I am engaged in an exercise in self-indulgence. I am trying to write the book that I would have liked to read back then and that gives me pleasure now in writing it.

It is perhaps worth emphasizing that I am not trying to provide an update of [17], or something like a *for dummies* version. Quite on the contrary, once the overlap in subject matter has been acknowledged, the two books could hardly be more different in all other respects. I approach things from the point of view of spectral theory, both with regard to the topics selected and the methods used, and the canonical systems themselves are the primary object of study. In [17], they make their first appearance about halfway through the book, as a tool to study the titular Hilbert spaces of entire functions. I do the opposite here: I use these spaces as a tool in inverse spectral theory and minimize their use otherwise. In particular, when I do start with the complex analytic methods in Chapter 4, I have one main goal: I want to establish as quickly as possible[2] that transfer matrices and their de Branges functions always come from an associated canonical system, and this will then again put everything firmly into a spectral theory framework.

I am now tempted to propose the view that this approach gives many results of [17] their proper setting. To give just one concrete example for this, consider the statement that one de Branges space will be isometrically contained in a second one if their de Branges functions are related by a transfer matrix. This is Theorem 34 of [17], and, since canonical systems have not been introduced at this point, it takes all of de Branges's considerable mathematical prowess to prove it. On the other hand, once the connection to canonical systems has been established, the result becomes a near triviality. It is then essentially Theorem 4.8(a) of this book, which has an extremely

2 This will not be very quick, though, as the result is deep.

easy proof. Another case in point is the Nevanlinna style parametrization of the measures that integrate functions from de Branges spaces correctly.

With not too much exaggeration, it can be said that I need exactly one (very) deep result from the complex analysis side of things here, and everything else will be spectral theory. This is de Branges's theorem on the total ordering, by inclusion, of the de Branges subspaces of a given space.

In any event, my goal here is to tell one continuous story, with everything explained in great detail. It should be immediately meaningful to anyone with some background in spectral theory already, but I hope that it will also be satisfying and feel coherent and complete to the novice in these areas.

Canonical systems are, in my opinion, not nearly as popular among spectral theorists currently as they deserve to be. From a mathematical point of view, they certainly occupy a central position among all second-order ordinary differential (and difference) operators. As we will discuss in detail in Chapter 5, arbitrary spectral data can be realized by a canonical system, and after a suitable normalization, these correspondences become bijections and even homeomorphisms between compact metric spaces. So canonical systems provide a unified and very elegant framework for spectral theory in one dimension, and these properties make them useful even if one was originally interested in more specialized operators.

Prerequisites. If I succeeded in giving the kind of presentation I had in mind, then the amount of detail should feel rather lavish to a moderately experienced mathematician. The book should also be accessible to graduate students, from what would correspond to the third year of their studies on at an average US university. I do assume that the reader can confidently handle classical real analysis (as presented in, for example, [25]), is thoroughly familiar with the spectral theorem for self-adjoint operators in Hilbert spaces, and knows complex analysis somewhat beyond the mere basics. Prior exposure to spectral theory in one dimension in a more classical setting (Schrödinger, Jacobi, or Dirac operators) would certainly be useful as background, but is not required.

Pronouns. I use pronouns in an extremely uncommon way (for mathematical writers). I use *you* to mean you (the reader), *I* to mean I (the author), and *we* means we as in you and I. Now strictly speaking this use of *we* is nonsensical when you and I will quite possibly never meet, but here I grant myself some artistic license, which seems harmless when, as discussed, in most of today's science writing the existence of a reader is already a literary fiction. I typically employ *we* to suggest a joint effort and encourage your active participation in something that I admittedly did before you read it, but it might help if you just pretend that we are doing it together right now and you are very much part of it.

Notes. Each chapter ends with a set of notes, which contain comments of various kinds. There are occasional references to the literature, but I am never concerned here, or anywhere else in this book, with assigning credit or establishing priority. The comments are meant to be useful to the reader wishing to study the material, not to the historian. I do not feel obliged to always mention all the work on a given subject that I am aware of, and no conclusions must be drawn from the absence of a reference. In any event, if these things matter to you, the rough summary is clear enough: next to the towering figure of de Branges, everyone else's contributions must inevitably look rather insignificant.

Highlights. I promised a few times that we will be discussing some very powerful and satisfying results about canonical systems here, and you might now want to know where exactly in this book these are, and I would then recommend the following gems, among many (I hope) others, to your attention: Theorem 3.5 and its reformulation, Corollary 3.6, which say that all limit circle endpoints are regular; de Branges's formula for the type of a transfer matrix (Theorem 4.26); obviously, the one-to-one correspondence between canonical systems and spectral data (Theorems 5.1, 5.2); the Nevanlinna type parametrization of measures that integrate functions from de Branges spaces correctly (Theorem 6.2); finally, I also hope that the results about the absolutely continuous spectrum will be found interesting, especially Theorems 7.12 and 7.18, though this betrays some vanity on my part since this last chapter is based mainly on my own recent work.

I welcome comments from readers on this book. If you have any, please write to christian.remling@ou.edu.

1 Basic definitions

1.1 Canonical systems

A *canonical system* is a differential equation of the form

$$Ju'(x) = -zH(x)u(x), \quad J = \begin{pmatrix} 0 & -1 \\ 1 & 0 \end{pmatrix}. \tag{1.1}$$

We consider these systems on open intervals $x \in (a, b)$, possibly unbounded, so $-\infty \leq a < b \leq \infty$, and we then make the following basic assumptions on the coefficient matrix H:

(1) $H(x) \in \mathbb{R}^{2\times 2}$;
(2) $H \in L^1_{\text{loc}}(a, b)$;
(3) $H(x) \geq 0$ for (Lebesgue) almost every $x \in (a, b)$.

What I am asking for in condition (2) is that the *entries* of H are locally integrable functions, and similar conventions will be used in the sequel when I talk about L^p conditions on vector or matrix valued functions. Condition (3) means that for almost every x, the matrix $H(x)$ is symmetric and $v^*H(x)v \geq 0$ for all $v \in \mathbb{C}^2$.

The z from (1.1) is sometimes referred to as the *spectral parameter;* but for now, all we need to know is that z is a complex number.

It is useful to make one more basic assumption from the outset, even though some things could be done without it. Namely, I also assume:

(4) $H(x) \neq 0$ for almost every $x \in (a, b)$.

If we had an interval (c, d) on which $H = 0$ almost everywhere, then the solutions would simply stay constant on (c, d), and removing the interval would not have any effect on the complement. If $H = 0$ on a more complicated positive measure set, then things are not as immediately obvious, but we could still use a transformation similar to the one discussed in Section 1.3 to remove such a set. More importantly, later results will confirm convincingly that making assumption (4) was the right choice.

The differential equation (1.1) has the general structure of an eigenvalue problem: namely, if we imagine a (formal) differential operator τ that acts on $\mathbb{C}^2$ valued functions u as $(\tau u)(x) = -H^{-1}(x)Ju'(x)$, then what (1.1) is asking for is that $\tau u = zu$. Of course, to make this rigorous, many questions have to be addressed much more carefully, for example: What are suitable domains for these operators, to make them self-adjoint? In fact, what Hilbert space do they act in? What if $H(x)$ fails to be invertible? We will investigate these in detail in the next chapter.

Canonical systems are of great mathematical interest because they are, in a precise sense, the most general class of symmetric second-order operators (and here order

https://doi.org/10.1515/9783110563238-001

really means "order of differentiation times dimension of the vector space the solutions take values in," and that indeed equals $1 \times 2 = 2$). This will be a major theme later in this book; in fact, we will show that after a suitable normalization, canonical systems on a half line $(0, \infty)$ are in one-to-one correspondence to all possible spectral data. This immediately gives great intrinsic interest to canonical systems, and it also makes them potentially very useful even if one is mainly interested in other, more specialized equations. For example, very basic operations on spectral data, such as applying an automorphism of $\mathbb{C}^+$ to a Titchmarsh–Weyl m function, will typically produce spectral data of a canonical system, not of a more specialized operator, even if that is what one started out with. This use of canonical systems is similar to the use of distributions in classical analysis. It allows one to perform operations with ease that otherwise would need laborious justification.

So, if any second-order problem is a canonical system, then it must in particular be possible to write the more classical second-order operators such as Schrödinger, Jacobi, Dirac, Sturm–Liouville operators, Krein strings as canonical systems, and this can indeed be done. The transformation for Schrödinger equations will be discussed in Section 1.3, and all the others are similar. I will later return to this topic, in Section 5.3, when I will show how to write a Jacobi difference equation as a canonical system. This will be useful there (but not strictly speaking necessary) to give a quick proof of the possibility of realizing large classes of spectral data by canonical systems, but of course it will also be of some independent interest.

We start the formal analysis of canonical systems by reviewing some basic results from the theory of ordinary differential equations. Since our coefficient matrix $H(x)$ can be quite irregular, the first thing we notice is that some interpretation of (1.1) is needed. In preparation for this, recall that a function $u : (a, b) \to \mathbb{C}^2$ is called (locally) *absolutely continuous*, and we then write $u \in AC$, if

$$u(x) = u(c) + \int_c^x f(t)\, dt$$

for some locally integrable function $f : (a, b) \to \mathbb{C}^2$ and $c \in (a, b)$; of course, as soon as we have this for a single c, it works for all $c \in (a, b)$. If $u \in AC$, then f is uniquely determined by u, up to a change of its values on a null set, and we call f the derivative of u. It follows from Lebesgue's differentiation theorem that u is differentiable in the classical sense at almost all $x \in (a, b)$, and $u'(x) = f(x)$. An alternative interpretation of absolutely continuous functions that is sometimes useful is provided by the theory of distributions: a locally integrable u will be in AC if and only if its distributional derivative $u' \in \mathcal{D}'$ is locally integrable.

We then call $u : (a, b) \to \mathbb{C}^2$ a *solution* of (1.1) if $u \in AC$ and (1.1) holds almost everywhere. This is the usual way to interpret such equations with singular coefficients. Note that if $H_1(x) = H_2(x)$ almost everywhere, then the two equations will have the same solutions in this sense. We will therefore not distinguish between coefficient functions that agree almost everywhere.

A clear indication that we have made a good definition is given by the fact that global existence and uniqueness of solutions generalizes to this setting. Let me state this as a formal theorem; however, for the proof, I refer the reader to the standard literature on ordinary differential equations, for example [12]. We will occasionally have to consider the inhomogeneous version of (1.1), so I will state the result in this generality.

Theorem 1.1. *Let $c \in (a,b)$ and let $f : (a,b) \to \mathbb{C}^2$ be locally integrable. Then, for any $v \in \mathbb{C}^2$, the initial value problem*

$$Ju' = -zHu + f, \quad u(c) = v \tag{1.2}$$

has a unique solution $u = u(x,z)$ on $x \in (a,b)$.

Moreover, $u(x,z)$ is jointly continuous on $(x,z) \in (a,b) \times \mathbb{C}$. For each fixed $x \in (a,b)$, the components of $u(x,z)$ are entire functions of $z \in \mathbb{C}$. The derivatives $u_n(x,z) := \partial^n u(x,z)/\partial z^n$ are themselves absolutely continuous functions of $x \in (a,b)$, and they solve the initial value problems that result from differentiating (1.2) *with respect to z formally, as follows:*

$$\begin{aligned}
Ju_1' &= -zHu_1 - Hu_0, \quad u_1(c) = 0,\\
Ju_2' &= -zHu_2 - 2Hu_1, \quad u_2(c) = 0,\\
&\dots\\
Ju_n' &= -zHu_n - nHu_{n-1}, \quad u_n(c) = 0.
\end{aligned}$$

If we take $f = 0$, then the uniqueness statement implies the well-known fact that the set of all solutions u to (1.1) is a two-dimensional vector space.

Transfer matrices are an extremely useful tool, even though they at first appear to be not much more than a piece of notation. They are defined as follows: the transfer matrix T is the matrix solution, taking values in $\mathbb{C}^{2\times 2}$, of the homogeneous equation (1.1) with the initial value $T(c) = 1$ (the identity matrix). So T will depend on $x, c \in (a,b)$, and $z \in \mathbb{C}$; I sometimes make all the arguments explicit and write $T(x,c;z)$, but equally frequently, it will be convenient to drop some or all of them. If the second argument is missing and the reader is in doubt, then $c = 0$ is the recommended guess.

Let us collect the basic properties of the transfer matrix. By the second part of the theorem, T is entire in z for fixed x, c. The columns of T solve (1.1), as a function of x; more generally, if $v \in \mathbb{C}^2$, then the unique solution with the initial value $u(c) = v$ is given by $u(x) = T(x,c;z)v$. So, as its name suggests, the transfer matrix updates values of solutions; more specifically, $T(x,c;z)$ gets us from c to x. The transfer matrix $T(x,c;z)$ is characterized by this property.

Since $J^{-1} = -J$, we have $T' = zJHT$, and the matrix JH has trace zero, as we can see from a calculation; in fact, this property is equivalent to H being symmetric. It thus follows that $\det T(x)$ is constant, and initially, $\det T = 1$, so this holds for all $x \in (a, b)$. In particular, T is invertible, and T^{-1} undoes what T did, and since T moved us from c to x, we obtain $T(x, c; z)^{-1} = T(c, x; z)$. We now see that T is also an absolutely continuous function of its second argument.

Theorem 1.2. *Fix $x, c \in (a, b)$, $x \geq c$. Then the transfer matrix $T(z) = T(x, c; z)$ has the following properties as a function of $z \in \mathbb{C}$: $T(z)$ is entire, $T(0) = 1$, $T(z) \in \mathrm{SL}(2, \mathbb{C})$, and if $z = t \in \mathbb{R}$, then $T(t) \in \mathrm{SL}(2, \mathbb{R})$. If $\operatorname{Im} z \geq 0$, then*

$$i(T^*(z)JT(z) - J) \geq 0. \tag{1.3}$$

Proof. We have proved the first few statements already, except for the claim that $T(0) = 1$, which is obvious because the equation becomes $T' = 0$ for $z = 0$. If $z = t$ is real, then $T(x, c; t)$ solves an initial value problem with real coefficients and real initial value, so has real entries itself.

It remains to establish (1.3), and this we do by the following calculation. As a preliminary, since $J^* = -J$, notice that taking adjoints in $JT' = -zHT$ gives $-T^{*\prime}J = -\overline{z}T^*H$. Write $z = t + iy$, $y \geq 0$, and consider

$$\frac{d}{dx}(T^*(x, c; z)JT(x, c; z)) = (\overline{z} - z)T^*HT = -2iyT^*HT.$$

Integration of this formula shows that the left-hand side of (1.3) equals $2y \int_c^x T^*HT\, ds$, and this is a positive definite matrix, as claimed. □

One of the major results that we will prove later will be that these conditions characterize transfer matrices: if a matrix function $T(z)$ is given that has the properties stated in the theorem, then there will be a canonical system on an interval (c, x) (in other words, there will be a coefficient function H on such an interval) such that $T(z) = T(x, c; z)$. Moreover, after a suitable normalization, the canonical system is uniquely determined by $T(z)$.

A variant of the calculation we did in the last proof produces another useful identity.

Theorem 1.3 (Constancy of the Wronskian). *We have*

$$T^t(x)JT(x) = J, \quad T(x) \equiv T(x, c; z).$$

As a consequence, the Wronskian $W(v, w) \equiv v^t(x)Jw(x)$ *of any two solutions v, w of the same equation $Ju' = -zHu$ is constant. Moreover, $W(v, w) = 0$ if and only if v, w are linearly dependent.*

Proof. The first identity is obvious from

$$(T^tJT)' = -(JT')^tT + T^tJT' = z(HT)^tT - zT^tHT = 0.$$

The final claim follows by just writing out $W(v, w) = v_2w_1 - v_1w_2$. □

Theorem 1.4 (Variation of constants). *The solution $u(x)$ to (1.2) can be obtained as*

$$u(x) = T(x)v - T(x)\int_c^x T^{-1}(t)Jf(t)\,dt, \quad T(x) \equiv T(x,c;z). \tag{1.4}$$

Another way of stating this is to say that the second term on the right-hand side of (1.4) is a solution of the inhomogeneous equation $Ju' = -zHu + f$. Recall in this context that the general solution to the inhomogeneous equation can be obtained as the general solution to the homogeneous equation $Ju' = -zHu$ plus a particular solution to the inhomogeneous equation.

Proof. Obviously, the u defined in (1.4) is well defined and absolutely continuous, it satisfies the initial condition at $x = c$, and it could of course be verified by a straightforward calculation that it solves the equation. However, it is also instructive to derive (1.4) more systematically, using the idea that gives the method its name. Namely, look for a solution of the form $u(x) = T(x)v(x)$. Plug this into the equation. This produces $JTv' = f$, so $v' = -T^{-1}Jf$, and now (1.4) follows by integrating this. □

1.2 Singular intervals

Definition 1.1. A point $x \in (a,b)$ is called *singular* if there are $\delta > 0$ and a vector $v \in \mathbb{R}^2$, $v \neq 0$, such that $H(t)v = 0$ for almost all $|t-x| < \delta$. A point $x \in (a,b)$ which is not singular is called *regular*.

The set S of singular points is open, and its connected components (c,d) are called *singular intervals*.

That S is open is of course immediate from the definition. We can thus write $S = \bigcup(c_j, d_j)$ as an at most countable union of disjoint open intervals, and these are the singular intervals introduced in the last part of the definition.

Now let $x \in (a,b)$ be a singular point. Since $H \geq 0$ and $H \neq 0$ almost everywhere on $(x-\delta, x+\delta)$, this function must be of the form $H(t) = h(t)P_\alpha$ on this neighborhood (almost everywhere, but recall that we identify coefficient functions that agree off a null set, so it is not necessary to always say this), for some $\alpha \in [0,\pi)$ and some function $h \in L^1(x-\delta, x+\delta)$, $h > 0$, and here

$$P_\alpha = \begin{pmatrix} \cos^2\alpha & \sin\alpha\cos\alpha \\ \sin\alpha\cos\alpha & \sin^2\alpha \end{pmatrix} = e_\alpha e_\alpha^*, \quad e_\alpha = \begin{pmatrix} \cos\alpha \\ \sin\alpha \end{pmatrix}$$

denotes the projection onto e_α. The same remarks apply to the whole singular interval that x is a part of. Let me introduce one more piece of terminology in this context.

Definition 1.2. The angle α is called the *type* of the singular interval (c,d).

Occasionally, I will also refer to the vector e_α as the type of the singular interval, which is somewhat inconsistent but convenient.

Definition 1.1 is very important, and its significance will become much clearer as we develop the material. We can get a preliminary idea of what this is about by solving (1.1) across a singular interval (c,d) of type α. So let us write $H(x) = h(x)P_\alpha \equiv h(x)P$ on (c,d). Since only the scalar h depends on x, it follows that any two matrices $JH(x)$, $JH(x')$ commute, so $u' = zJHu$ is solved by the matrix exponential

$$u(x) = e^{z(\int_c^x h(t)\,dt)JP}u(c).$$

Expand this into its power series. Now $(JP)^2 = JPJP = 0$; in fact, $PJP = 0$ since J acts as a rotation by 90 degrees. This means that the series for the exponential terminates after the first two terms, and thus

$$u(x) = \left(1 + z\left(\int_c^x h(t)\,dt\right)JP\right)u(c)$$

(or forget all this and just confirm that this u solves (1.1)). In particular, $u(d) = (1 + zJH)u(c)$, with $H = \int_c^d H(x)\,dx$, but this is just the same as doing one recursion step in the difference equation analog

$$J(u_{n+1} - u_n) = -zH_n u_n$$

of (1.1), if here $H_n = H$.

From a more abstract point of view, the unexpected property of the transfer matrix $T = 1 + zJH$ across a singular interval is its polynomial dependence on z, of degree 1 at that. Again, this is what recursions produce, while differential equations would typically lead to more complicated functions of z.

So the upshot of all this is that a canonical system across a singular interval mimics a difference equation. The ability of a canonical system to do this is crucial. I already mentioned the result that any spectral data whatsoever can be realized by a canonical system, and these spectral data could come from a (Jacobi) difference equation, so the option to simulate these is needed.

As we will discover later, singular intervals are also used to implement boundary conditions, and they are responsible for the multi-valued part of the relations that we will associate with canonical systems.

1.3 Transformations

Canonical systems have the property that their general form is preserved under any monotone change of variable. (Schrödinger equations for example do not have this property; in fact, no non-identity transformation of the variable will yield a

Schrödinger equation again.) Let us make this more explicit. Let $p \in L^1_{\mathrm{loc}}(a,b)$, $p > 0$, fix a $c \in (a,b)$, and introduce the new variable

$$X = \int_c^x p(t)\,dt. \tag{1.5}$$

This defines a bijection $x \mapsto X$ from (a,b) to a new interval (A,B), and both $x(X)$ and $X(x)$ are absolutely continuous. Define $H_1(X) = (1/p(x))H(x)$, with x and X related by (1.5). Then $H_1(X)$ also satisfies the basic assumptions on the coefficient function of a canonical system; note in particular that the substitution rule shows that $H_1 \in L^1_{\mathrm{loc}}(A,B)$.

Theorem 1.5. *For an absolutely continuous function $u : (a,b) \to \mathbb{C}^2$, define $U(X) = u(x(X))$. Then u solves $J(du/dx) = -zHu$ if and only if U solves $J(dU/dX) = -zH_1U$, and*

$$\int_x^y u^*(t)H(t)u(t)\,dt = \int_X^Y U^*(T)H_1(T)U(T)\,dT$$

for every interval $(x,y) \subseteq (a,b)$.

Proof. This is easy to verify, by direct calculation. □

So it makes sense to think of H and H_1 as different representations of equivalent canonical systems, and then it might seem useful to pick a distinguished representative from each equivalence class. This is usually done by asking that $\operatorname{tr} H(x) = 1$, and such a canonical system is called *trace normed*. Since $p = \operatorname{tr} H$ is an admissible choice in (1.5), Theorem 1.5 shows that every canonical system is equivalent to a unique, up to a shift of the basic interval, trace normed system in the sense that the two can be transformed into each other by a change of variable as in (1.5).

One could focus on trace normed systems exclusively, but in some situations, the extra flexibility of an unrestricted H is useful, so we will develop the theory in this generality.

A completely different but equally useful class of transformations is given by

$$H_A(x) = A^{-1t}H(x)A^{-1}, \quad A \in SL(2,\mathbb{R}).$$

Theorem 1.6. *With H_A defined as above, the transfer matrices satisfy $T_A(x,c;z) = AT(x,c;z)A^{-1}$.*

Proof. Clearly, $T_A(c,c;z) = 1$ has the correct initial value, so we only need to verify that T_A solves the equation $JT'_A = -zH_AT_A$. We compute

$$JT'_A = -JAJJT'A^{-1} = zJAJHTA^{-1} = zJAJHA^{-1}T_A,$$

and this is of the desired form because $JAJ = -A^{-1t}$. □

Now let me discuss still another type of transformation, not between canonical systems themselves this time, but from a *Schrödinger equation*

$$-y''(x) + V(x)y(x) = zy(x), \quad x \in (a,b), \tag{1.6}$$

to a canonical system. Here we make the usual assumption that $V \in L^1_{loc}(a,b)$. In the canonical system, the spectral parameter z multiplies the whole coefficient function $H(x)$. To give (1.6) a similar property, it is natural to try variation of constants about $z = 0$. To prepare for this, we introduce the transfer matrix formalism for (1.6) also. First of all, we rewrite (1.6) in matrix form, as follows:

$$Y' = \begin{pmatrix} 0 & V(x) \\ 1 & 0 \end{pmatrix} Y + z \begin{pmatrix} 0 & -1 \\ 0 & 0 \end{pmatrix} Y \equiv (A + zB)Y. \tag{1.7}$$

This is set up so that $Y = (y', y)^t$ solves (1.7) if and only if $y = Y_2$ solves the original equation (1.6). Now let $T_0(x)$ be the 2×2 matrix solution of $T_0' = AT_0$, $T_0(0) = 1$. In other words,

$$T_0(x) = \begin{pmatrix} p'(x) & q'(x) \\ p(x) & q(x) \end{pmatrix},$$

and here p, q both solve (1.6) for $z = 0$, and $p(0) = q'(0) = 0$, $p'(0) = q(0) = 1$. Note that again $\det T_0(x) = 1$ for all x.

Now, given a solution Y of (1.7) (for general $z \in \mathbb{C}$), introduce u by writing $Y(x) = T_0(x)u(x)$; this is the variation of constants method. Then a quick calculation shows that u solves the canonical system

$$Ju' = -zHu, \quad H = \begin{pmatrix} p^2 & pq \\ pq & q^2 \end{pmatrix}.$$

These steps can be taken in reverse, so it is also true that if u solves this, then $Y := T_0 u$ will solve (1.7). So the two equations are equivalent in this sense, and we have succeeded in rewriting the Schrödinger equation as a canonical system. The transformation also preserves other quantities of importance in spectral theory, such as m functions and L^2 norms, but we have not introduced these yet, and I leave the matter at that. As I mentioned above, we will briefly return to these topics in Section 5.3, in a different context, and then again in Sections 6.1, 6.4, and 7.1.

This canonical system that is a Schrödinger equation rewritten also suggests a basic intuition about general canonical systems that is sometimes useful. First of all, it is certainly always true that $H(x)$ itself very explicitly contains all the necessary information about the solutions at specifically $z = 0$: this is just a convoluted way of saying that these solutions are constant. We could then think of the canonical system at a general z as a variation of constants of sorts (but now of an unknown or even fictitious equation) about $z = 0$, at which value everything is explicit. In other words, one could argue that a canonical system does not really have a coefficient function, not in the same way at least as the classical equations; instead, we are explicitly given what we need to know to work out norms of solutions at $z = 0$ in the Hilbert space $L^2_H(a, b)$, which will be the subject of the next section.

1.4 The Hilbert space L^2_H

Later, when we study self-adjoint realizations of (1.1) and their spectral theory, we will need a Hilbert space that these act in, and the appropriate space for this is $L^2_H(a, b)$, defined as follows. Suppose that $H(x)$ satisfies the assumptions listed at the beginning of Section 1.1. Let

$$\mathcal{L} = \left\{ f : (a, b) \to \mathbb{C}^2 : f \text{ (Borel) measurable}, \int_a^b f^*(x)H(x)f(x)\, dx < \infty \right\},$$

write $\|f\| = (\int_a^b f^* Hf\, dx)^{1/2}$ for $f \in \mathcal{L}$, and define $L^2_H(a, b)$ as the quotient $\mathcal{L}/\mathcal{N}$, with $\mathcal{N} = \{f \in \mathcal{L} : \|f\| = 0\}$.

Of course, this is just the usual procedure to define L^p spaces, except perhaps for the extra feature that our functions take values in $\mathbb{C}^2$ rather than $\mathbb{C}$. There is not much to say about this really. Everything works the same way as in the classical setting and L^2_H is a separable, infinite-dimensional Hilbert space. In fact, the map

$$V : L^2_H(a, b) \to L^2_I(a, b), \quad (Vf)(x) = H^{1/2}(x)f(x),$$

with $H^{1/2}(x)$ defined as the unique positive square root of $H(x)$, provides an embedding of L^2_H into L^2_I, with I being the 2×2 identity matrix. Since there is an obvious identification $L^2_I(a, b) \cong L^2(a, b) \oplus L^2(a, b)$, this reduces matters very explicitly to the classical L^2 spaces, and this map is sometimes a useful device.

There is perhaps one feature of L^2_H that is worth commenting on more explicitly: since $f^* Hf = (H^{1/2}f)^* H^{1/2}f$, functions f, g will represent the same element of L^2_H if and only if $H(x)f(x) = H(x)g(x)$ for (Lebesgue) almost every x, but if $H(x)$ has a kernel for some x, so $H(x)v(x) = 0$, $v(x) \neq 0$, then this just says that $f(x) - g(x) = c(x)v(x)$ for those x. In particular, it is entirely possible for an $f \in L^2_H$ to have several continuous representatives that are not equal as functions.

Finally, the following simple fact will be used frequently.

Lemma 1.7. *If $f \in L^2_H(a, b)$, then $Hf \in L^1_{\text{loc}}(a, b)$.*

Proof. This follows from the Cauchy–Schwarz inequality: take the positive square root $H^{1/2}(x)$ and write $Hf = H^{1/2}(H^{1/2}f)$. Both factors are now in (matrix and vector valued, respectively) $L^2_{\text{loc}}(a, b)$. □

1.5 Notes

Section 1.1. There is no agreement in the literature on how exactly to write a canonical system or what to call it. You will also find these equations referred to as (linear) Hamiltonian systems; de Branges, in his work, does not call them anything.

Some authors drop the minus sign on the right-hand side, or take transposes, or both; de Branges, in addition to this, also passes to the integrated form and then replaces $H(x)\,dx$ by a general continuous measure $d\mu(x)$. This is not really more general because one can again use a transformation of the independent variable, as discussed in Section 1.3, to make the measure absolutely continuous anyway. Obviously, none of this really matters. The form I chose is convenient because it will lead to the formula $m(z) = f(0, z)$ for the m function later.

Section 1.2. The lack of standardization continues. I do not see any reason to deviate from de Branges's terminology here, which seems straightforward and appropriate; other authors refer to singular intervals as *H-indivisible intervals*.

2 Symmetric and self-adjoint relations

2.1 The minimal and maximal relations of a canonical system

We now look in detail at the question of how a canonical system

$$Ju'(x) = -zH(x)u(x), \quad x \in (a,b)$$

generates self-adjoint operators on the Hilbert space $L^2_H(a,b)$. We already know that we would like these operators to act as $-H^{-1}Jf'$ on functions f from their domains, and we have also already observed that there is no hope to make this formula work in general since $H(x)$ need not be invertible. We can avoid this problem by moving H back to the other side, as follows: suppose we have a pair $f, g \in L^2_H$, and we want to think of $g = \tau f$ as the result of an application of the operator we are trying to construct to f. This can then formally be written as

$$Jf'(x) = -H(x)g(x), \tag{2.1}$$

and we will now make (2.1) our starting point. This condition will in general not define an operator, but only a *relation*.

Definition 2.1. Let $\mathcal{H}$ be a Hilbert space. A *relation* is a linear subspace of $\mathcal{H} \oplus \mathcal{H}$.

Linear operators T then become special relations, after identifying them with their graphs $\{(x, Tx)\}$. Conversely, relations are best thought of as being the same thing as operators except that an $f \in \mathcal{H}$ can have several images. A relation $\mathcal{T}$ is an operator if $(f, g_1), (f, g_2) \in \mathcal{T}$ implies that $g_1 = g_2$. By linearity, this is equivalent to the condition that $(0, g) \in \mathcal{T}$ only if $g = 0$.

We now define the *maximal relation* $\mathcal{T}$ of a canonical system as the collection of all pairs (f, g) for which (2.1) (can be easily interpreted and) holds. More precisely, we set

$$\mathcal{T} = \{(f,g) : f, g \in L^2_H(a,b), f \text{ has a representative } f_0 \in AC \\ \text{such that } Jf_0'(x) = -H(x)g(x) \text{ for a.e. } x \in (a,b)\}. \tag{2.2}$$

This clearly does define a linear subspace or, synonymously, a relation.

As I already pointed out in the first chapter, an $f \in L^2_H$ can have several continuous representatives, so we cannot really expect f_0 to be uniquely determined by f (and it is not in general, as we will see soon). Therefore, it pays to be very careful about the distinction between Hilbert space elements (= equivalence classes of functions) and functions here.

While the f_0 from (2.2) need not be determined by f, what is true is that f_0 is determined by the pair (f, g), if we rule out the trivial scenario where (a, b) is just a single

https://doi.org/10.1515/9783110563238-002

singular interval. So we make the following **basic assumption:**

The interval (a, b) contains at least one regular point.

This will be in force throughout from now on, unless explicitly stated otherwise. If (a, b) is a single singular interval, then everything can be worked out explicitly and we do not need any general theory at all. More to the point, many of our subsequent results need modification in this case, so it is much more convenient to just exclude this trivial scenario from the development of the general theory.

Let us now return to my claim that if $(f, g) \in \mathcal{T}$, then the f_0 from (2.2) is uniquely determined. Indeed, integration of $Jf_0' = -Hg$ shows that

$$f_0(x) = f_0(c) + J \int_c^x H(t)g(t)\,dt;$$

note that here $H(t)g(t)$ can only change on a null set if we choose a different representative of g, so the integral is determined by the Hilbert space element $g \in L_H^2$. It follows that if we had two such representatives of f, then they could only differ by a constant function v, but then we must have $H(x)v = 0$ for almost every x, or they would not represent the same Hilbert space element. If $v \neq 0$ here, this means that (a, b) is a singular interval, which we explicitly ruled out.

This general fact is very important for the developments of this chapter, and the reader should keep it firmly in mind. For further emphasis, let me state it as a formal result.

Lemma 2.1. *An element $(f, g) \in \mathcal{T}$ of the maximal relation uniquely determines an absolutely continuous function $f_0 : (a, b) \to \mathbb{C}^2$ with the following two additional properties:*
(1) *$f_0 \in L_H^2(a, b)$, and it represents the element f;*
(2) *$Jf_0' = -Hg$.*

We will usually employ this careful notation in the sequel also. In other words, subscript 0 as in f_0 means that we are working with the representative of f that is determined (not just by f itself, but) by the pair $(f, g) \in \mathcal{T}$.

Let us now see an example where f_0 indeed is not determined by just f; this will also confirm that we really need relations, not just operators. Let us look at the case that we just dismissed, where (a, b) is a singular interval itself. I will take $a = 0, b = 1$, $H(x) = \left(\begin{smallmatrix} 1 & 0 \\ 0 & 0 \end{smallmatrix}\right)$.

Then $(f, g) \in \mathcal{T}$ if and only if $f_{0,2}' = g_1, f_{0,1}' = 0$. Only the first component of a function matters from the point of view of the Hilbert space. Thus we can take any (absolutely continuous) function whatsoever as our $f_{0,2}$, and it follows that $g \in L_H^2(0, 1)$ is arbitrary. Obviously, $f_{0,1}$ must be constant. So, writing e_1 for the first unit vector, we have found

$$\mathcal{T} = L(e_1) \oplus L_H^2(0, 1) = \{(f, g) : f(x) = ce_1, g \in L_H^2(0, 1)\}.$$

We now state a few definitions for relations in general that, for the most part, adapt operator concepts in an obvious way to the new setting.

Definition 2.2. Let $\mathcal{T} \subseteq \mathcal{H} \oplus \mathcal{H}$ be a relation. We call $\mathcal{T}$ *closed* if $\mathcal{T}$ is a closed subspace of $\mathcal{H} \oplus \mathcal{H}$. Similarly, the *closure* $\overline{\mathcal{T}}$ of $\mathcal{T}$ is just that, the closure of the subspace $\mathcal{T}$.

We define the *domain, range, null space* (or *kernel*), and *multi-valued part* of $\mathcal{T}$ as follows:

$$\begin{aligned} D(\mathcal{T}) &= \{f \in \mathcal{H} : (f,g) \in \mathcal{T} \text{ for some } g\}, \\ R(\mathcal{T}) &= \{g \in \mathcal{H} : (f,g) \in \mathcal{T} \text{ for some } f\}, \\ N(\mathcal{T}) &= \{f \in \mathcal{H} : (f,0) \in \mathcal{T}\}, \\ \mathcal{T}(0) &= \{g \in \mathcal{H} : (0,g) \in \mathcal{T}\}. \end{aligned}$$

The *inverse* of $\mathcal{T}$ is the relation

$$\mathcal{T}^{-1} = \{(g,f) : (f,g) \in \mathcal{T}\},$$

and the *adjoint* of $\mathcal{T}$ is defined as

$$\mathcal{T}^* = \{(h,k) : \langle h,g\rangle = \langle k,f\rangle \text{ for all } (f,g) \in \mathcal{T}\}.$$

We call $\mathcal{T}$ *symmetric* if $\mathcal{T} \subseteq \mathcal{T}^*$ and *self-adjoint* if $\mathcal{T} = \mathcal{T}^*$.

So, unlike operators, relations always have closures, inverses, and unique adjoints, and this can make them a rather useful tool, even if one is interested in operators exclusively. The role relations can play in these situations is perhaps similar to the use of distributions in classical analysis.

Let us now try to find the adjoint $\mathcal{T}_0 := \mathcal{T}^*$ of the maximal relation $\mathcal{T}$ of a canonical system. To do this, we introduce a third relation,

$$\mathcal{T}_{00} = \{(f,g) \in \mathcal{T} : f_0(x) \text{ has compact support in } (a,b)\}.$$

It is easy to see that $\mathcal{T}_{00} \subseteq \mathcal{T}^*$. Indeed, fix $(f,g) \in \mathcal{T}_{00}$ and let $(h,k) \in \mathcal{T}$ be arbitrary. We must show that then $\langle f,k\rangle = \langle g,h\rangle$ or, if we write it out,

$$\int_a^b f^*(x)H(x)k(x)\,dx = \int_a^b g^*(x)H(x)h(x)\,dx.$$

Now plug $Hk = -Jh_0'$, $Hg = -Jf_0'$ into this, and then we must show that

$$\int_a^b (f_0^*(x)Jh_0'(x) + f_0^{*\prime}(x)Jh_0(x))\,dx = 0,$$

which is obvious because f_0 is zero near both a and b.

We now come to the crucial step.

Proposition 2.2. *We have $\mathcal{T}_{00}^* \subseteq \mathcal{T}$.*

Proof. Let $(f, g) \in \mathcal{T}_{00}^*$ and define a function f_1 by

$$f_1(x) = J \int_c^x H(t)g(t)\, dt,$$

for some fixed $c \in (a, b)$. Then f_1 is absolutely continuous and $Jf_1' = -Hg$, but of course there is no guarantee at this point that f_1 will be square integrable and represent a Hilbert space element.

Let $(h, k) \in \mathcal{T}_{00}$ be arbitrary. An integration by parts then shows that

$$\begin{aligned}\langle h, g\rangle &= \int_a^b h_0^*(x)H(x)g(x)\, dx = -\int_a^b h_0^*(x)Jf_1'(x)\, dx \\ &= \int_a^b k^*(x)H(x)f_1(x)\, dx.\end{aligned}$$

Note that h_0 and Hk have compact support, so the fact that f_1 may fail to lie in L_H^2 cannot make the last integral divergent. For the same reason, the integration by parts does not contribute boundary terms.

On the other hand, $\langle h, g\rangle = \langle k, f\rangle = \int_a^b k^* Hf$, so we conclude that

$$\int_a^b k^*(x)H(x)(f_1(x) - f(x))\, dx = 0 \quad \text{for all } k \in R(\mathcal{T}_{00}). \tag{2.3}$$

Observe now that a $k \in L_H^2(a, b)$ will be in $R(\mathcal{T}_{00})$ precisely if it satisfies the following two conditions: (1) Hk is compactly supported; (2) $\int_a^b Hk = 0$. (Since Hk is locally integrable and compactly supported, this integral is defined.) Let us denote by X the linear subspace of L_H^2 defined by condition (1) and consider the functionals

$$F_j(k) = e_j^* \int_a^b H(x)k(x)\, dx, \quad F(k) = \int_a^b (f_1(x) - f(x))^* H(x)k(x)\, dx$$

on X. What we just established in (2.3) can then be rephrased as the statement that if $F_1(k) = F_2(k) = 0$ for a $k \in X$, then $F(k) = 0$. We now use the linear algebra fact, established in Lemma 2.3 below, that then F must be a linear combination of F_1, F_2. So there is a vector $v \in \mathbb{C}^2$ such that

$$\int_a^b (f_1(x) - f(x) - v)^* H(x)k(x)\, dx = 0$$

for all $k \in X$. Since $f_1 - f - v$ is locally in L^2_H, this is only possible if $H(x)(f_1(x) - f(x) - v) = 0$ almost everywhere. We have shown that f has the absolutely continuous representative $f_1(x) - v$, and $J(f_1 - v)' = -Hg$ by construction of f_1. This says that $(f,g) \in \mathcal{T}$, as claimed. □

Lemma 2.3. *Let $F_1, \dots, F_n, F : X \to \mathbb{C}$ be linear functionals on a vector space X and assume that $\bigcap N(F_j) \subseteq N(F)$. Then F is a linear combination of the F_j.*

Proof. We proceed by induction on n. The case $n = 1$ is clear because the null space of a non-zero functional has codimension 1, so determines the functional up to a constant factor.

Now assume the claim for n and consider $n + 1$. If $N(F_{n+1}) \supseteq \bigcap_{j=1}^n N(F_j)$, then $F = \sum_{j=1}^n a_j F_j$, by the induction hypothesis. In the other case, we can pick an $x \in N(F_j)$, $j = 1, \dots, n$, $x \notin N(F_{n+1})$, so $F_{n+1}(x) \neq 0$, and let us make $F_{n+1}(x) = 1$. Consider the new functional

$$G(y) = F(y) - F(x)F_{n+1}(y).$$

If $y \in N(F_j)$, $j = 1, \dots, n$, then $z := y - F_{n+1}(y)x \in N(F_j)$ also, since we chose $x \in N(F_j)$. Moreover, $F_{n+1}(z) = 0$, so $z \in \bigcap_{j=1}^{n+1} N(F_j)$. Thus $z \in N(F)$, by assumption, and this implies that $G(y) = 0$. We have shown that $\bigcap_{j=1}^n N(F_j) \subseteq N(G)$, so $G = \sum_{j=1}^n a_j F_j$ by the induction hypothesis and hence

$$F(y) = \sum_{j=1}^{n+1} a_j F_j(y), \quad a_{n+1} = F(x),$$

as required. □

We are now almost ready to give our first description of the *minimal relation* $\mathcal{T}_0 = \mathcal{T}^*$. We need a few more easy general facts about relations.

Proposition 2.4. *Let $\mathcal{T} \subseteq \mathcal{H} \oplus \mathcal{H}$ be a relation. Then:*
(a) *$\mathcal{T}^*$ is closed;*
(b) *$\mathcal{T}^{**} = \overline{\mathcal{T}}$;*
(c) *$\overline{\mathcal{T}}^* = \mathcal{T}^*$.*

Proof. All three statements follow immediately after noticing that the adjoint is essentially an orthogonal complement. More precisely, if S denotes the unitary map $S(x,y) = (y, -x)$ on $\mathcal{H} \oplus \mathcal{H}$, then $\mathcal{T}^* = S\mathcal{T}^\perp = (S\mathcal{T})^\perp$. □

Theorem 2.5. (a) *The maximal relation $\mathcal{T}$ is closed.*
(b) *The minimal relation $\mathcal{T}_0 = \mathcal{T}^*$ is closed and symmetric, and $\mathcal{T}_0 = \overline{\mathcal{T}_{00}}$, $\mathcal{T}_0^* = \mathcal{T}$.*

Proof. (a) Assume that $(f_n, g_n) \in \mathcal{T}$, $(f_n, g_n) \to (f,g) \in \mathcal{H} \oplus \mathcal{H}$. We must show that then $(f,g) \in \mathcal{T}$ also.

By passing to a subsequence, we may assume that $H(x)f_{n,0}(x) \to H(x)f(x)$ pointwise almost everywhere, for the representatives from Lemma 2.1. Notice also that for each fixed $x \in (a, b)$, the sequence $f_{n,0}(x)$ must be bounded. This follows because the derivatives $f'_{n,0} = JHg_n$ are bounded in $L^1(c, d)$ for any compact subset $[c, d] \subseteq (a, b)$, so if $|e \cdot f_{n,0}(x)|$ were large for some direction $e \in \mathbb{C}^2$, $\|e\| = 1$, then the same would be true on any compact subset of (a, b), but that would make the norm of f_n large, since (a, b) is not a single singular interval and thus $H(x)$ cannot annihilate e everywhere.

So we can choose our subsequence such that, in addition, $f_{n,0}(c) \to v$, for a fixed $c \in (a, b)$ that we chose in advance. Now we can simply pass to the (pointwise) limit in

$$f_{n,0}(x) = f_{n,0}(c) + J \int_c^x H(t)g_n(t)\, dt.$$

We see that $f_{n,0}(x)$ itself converges (not just after applying $H(x)$), and, as we noted earlier, the limit will represent f. We have found a representative of f that is absolutely continuous and satisfies $Jf' = -Hg$, so $(f, g) \in \mathcal{T}$, as desired.

(b) Obviously, $\mathcal{T}_0$ is closed, being an adjoint, and the symmetry will follow from the two identities stated at the end, so it suffices to prove these. We already showed that $\mathcal{T}_{00} \subseteq \mathcal{T}^*$ and (Proposition 2.2) $\mathcal{T}_{00}^* \subseteq \mathcal{T}$, and taking adjoints in the second inclusion gives $\mathcal{T}_{00}^{**} \supseteq \mathcal{T}^*$. If we take closures in the first inclusion and use Proposition 2.4(a), (b), then this implies that $\overline{\mathcal{T}_{00}} = \mathcal{T}^* = \mathcal{T}_0$. Taking adjoints one more time then produces the final claim. □

We can give a more explicit description of the minimal relation $\mathcal{T}_0$. In fact, part of the result at least can be guessed already: $\mathcal{T}_0$ can be obtained by taking the closure of $\mathcal{T}_{00}$, which was defined as those elements of the maximal relation for which f_0 has compact support. Now it seems reasonable to assume that the effect of the closure will be a condition that says that f_0 has to be zero at the boundary of our interval (a, b). This is literally true at *regular endpoints.*

Definition 2.3. The endpoint a is called *regular* if $H \in L^1(a, c)$ for some (and then all) $c \in (a, b)$, and similarly for b.

This has nothing to do with our earlier classification into singular and regular points in Definition 1.1, and it does not interfere with it either since the latter refers to points $x \in (a, b)$ while Definition 2.3 applies to the endpoints.

Definition 2.3 differs slightly from its traditional version, which would also require a regular endpoint to be finite; in my version, $a = -\infty$, $b = \infty$ can be regular endpoints. This version works really much better for canonical systems. Infinite regular endpoints behave exactly the same way as those lying in $\mathbb{R}$. In fact, infinite endpoints can always be made finite by a transformation of the type discussed in Theorem 1.5; exactly the regular points are mapped to regular points again under such a transformation.

Note, however, that this is a peculiar feature of canonical systems, and it would not be a good idea to allow regular endpoints to be infinite for other equations. We want a definition that gives us Lemma 2.6 below, and that would clearly not work for (say) Schrödinger equations $-u'' + Vu = zu$ with $V \in L^1$ on a half line (not even if $V \equiv 0$).

Lemma 2.6. *If a is regular, then for any $(f,g) \in \mathcal{T}$, the representative f_0 has a continuous extension to $[a,b)$, and in fact $f_0 \in AC[a,b)$. Moreover, solutions to homogeneous equations $Ju' = -zHu$ have the same properties.*

The analogous result holds if b is regular.

We already knew that $f_0 \in AC(a,b)$, and this means that $f_0(x) = f_0(c) + \int_c^x h(t)\,dt$ for some $h \in L^1_{\text{loc}}(a,b)$. The claim that $f_0 \in AC[a,b)$ now makes the additional statement that $h \in L^1(a,c)$ here, for $c \in (a,b)$. This implies that f_0 has a continuous extension to $x = a$, but does not follow from this property, as you can see from examples such as $f(x) = x\sin(1/x)$.

No modifications to these statements are necessary in the case $a = -\infty$, if we give the extended interval $[a,c) = [-\infty,c)$ its obvious topology.

Proof. The Cauchy–Schwarz inequality shows that for any $g \in L^2_H(a,c)$, we have $Hg = H^{1/2}H^{1/2}g \in L^1(a,c)$, so the claims on f_0 are obvious from

$$f_0(x) = f_0(c) + J\int_c^x H(t)g(t)\,dt.$$

As for solutions u of $Ju' = -zHu$, we just apply basic ordinary differential equation theory, as summarized in Theorem 1.1 of Chapter 1, to the initial value problem $u(a) = v$ for general $v \in \mathbb{C}^2$ to confirm that u is absolutely continuous on $[a,b)$.

Possibly the reader has now developed some anxiety about the validity of this last argument in the case $a = -\infty$, but this is unnecessary. The initial value problem $Ju' = -zHu$, $u(-\infty) = v$ can be rewritten as an integral equation $u(x) = v + zJ\int_{-\infty}^x H(t)u(t)\,dt$, and if $H \in L^1(-\infty,c)$, then this can be solved exactly as on a bounded interval by Picard iteration. Or, as discussed above, we run a transformation to make $a = -\infty$ a finite endpoint $A \in \mathbb{R}$ and then avoid these issues altogether. □

Lemma 2.7. *Let $(c,d) \subseteq (a,b)$ and suppose that neither (a,c) nor (d,b) is a non-empty interval that is contained in a single singular interval. Let $(h,k) \in \mathcal{T}_{(c,d)}$. Then there is an $(f,g) \in \mathcal{T}$ with $f_0 = h_0$ on (c,d), $f_0(x) = 0$ for $x \in (a,c)$ near a and $x \in (d,b)$ near b.*

Here, $c = a$ or $d = b$ (or both, but then everything becomes trivial) is allowed, and in fact this will be our first application of the lemma.

Proof. Let's say $d < b$. Since d is a regular endpoint of (c,d), Lemma 2.6 applies and we conclude that h_0 is absolutely continuous on $(c,d]$. We must provide an absolutely continuous function f_0 on $[d,b)$ such that $Jf_0' = -Hg$ for some $g \in L^2_H(d,b)$, and with $f_0(d) = h_0(d)$, to make the function absolutely continuous overall, when we glue together the two pieces. We also want $f_0(x) = 0$ at all large x, and again, it is enough to reach this value once because then we can tack on the zero function from that point on. So we want $f_0(d) = h_0(d)$ and $f_0(t) = 0$, and here we take the precaution of taking $t \in (d,b)$ so large that (d,t) is not contained in a singular interval. If we say all this in terms of g, then what we want to do now is find a $g \in L^2_H(d,t)$ such that the function f_0 defined by

$$f_0(x) = h_0(d) + J\int_d^x H(s)g(s)\,ds$$

satisfies $f_0(t) = 0$. This will certainly work if the linear map

$$F : L^2_H(d,t) \to \mathbb{C}^2, \quad F(g) = \int_d^t H(s)g(s)\,ds$$

is surjective, and it is easy to see that this will be the case if (d,t) is not contained in a singular interval because then the range of $H(x)$ cannot identically be equal to a fixed one-dimensional subspace of $\mathbb{C}^2$.

Finally, if also $c > a$, then we apply the same procedure to the left of (c,d). □

Theorem 2.8. *Let $(f,g),(h,k) \in \mathcal{T}$. Then $f_0^*(x)Jh_0(x)$ approaches limits as either $x \to a+$ or $x \to b-$. Moreover,*

$$\langle g,h\rangle - \langle f,k\rangle = f_0^* J h_0 |_a^b.$$

We will use the slightly deceptive, but convenient notation $(f_0^*Jh_0)(a)$ and $(f_0^*Jh_0)(b)$ for these limits. In fact, the final claim of Theorem 2.8 already employs this notation or something very similar.

If an endpoint (let's say, a) is regular, then the existence of these becomes an immediate consequence of Lemma 2.6, and in this case, $(f_0^*Jh_0)(a) = f_0^*(a)Jh_0(a)$. Here, we use the suggestive notation $f_0(a)$, $h_0(a)$ for the continuous extensions of these functions to $x = a$.

Proof. Both statements follow from the following calculation:

$$\begin{aligned}
\langle g,h\rangle - \langle f,k\rangle &= \lim_{\substack{\alpha\to a+\\ \beta\to b-}} \int_\alpha^\beta (g^*(x)H(x)h_0(x) - f_0^*(x)H(x)k(x))\,dx \\
&= \lim_{\substack{\alpha\to a+\\ \beta\to b-}} \int_\alpha^\beta (f_0^{*\prime}(x)Jh_0(x) + f_0^*(x)Jh_0'(x))\,dx \\
&= \lim_{\substack{\alpha\to a+\\ \beta\to b-}} f_0^* J h_0 |_\alpha^\beta.
\end{aligned}$$
□

We are now in a position to give the promised description of the minimal relation.

Theorem 2.9. *Let $(f,g) \in \mathcal{T}$. Then $(f,g) \in \mathcal{T}_0$ if and only if*

$$(f_0^*Jh_0)(a) = (f_0^*Jh_0)(b) = 0 \quad \textit{for all } (h,k) \in \mathcal{T}. \tag{2.4}$$

If a is a regular endpoint, then the corresponding condition simplifies to $f_0(a) = 0$, *and similarly at b.*

Proof. Let us denote the relation defined by (2.4) by $\mathcal{T}_1$. Then the identity from Theorem 2.8 implies that $\mathcal{T}_1 \subseteq \mathcal{T}^* = \mathcal{T}_0$. Conversely, if $(f,g) \in \mathcal{T}$, but $(f,g) \notin \mathcal{T}_1$, then there is an $(h,k) \in \mathcal{T}$ such that $f_0^* J h_0$ is non-zero at one endpoint at least, let's say at a. But now we can use Lemma 2.7 to modify h_0 so that the new function is zero near b, and h_0 is left untouched near a. This modification corresponds to a new element $(h_1,k_1) \in \mathcal{T}$, and Theorem 2.8 now shows that $\langle f,k_1\rangle \neq \langle g,h_1\rangle$. Hence $(f,g) \notin \mathcal{T}^* = \mathcal{T}_0$.

If a is regular, observe that $\{h_0(a) : (h,k) \in \mathcal{T}\} = \mathbb{C}^2$. This follows again from Lemma 2.7, by starting with an element $(h_1,k_1) \in \mathcal{T}_{(a,c)}$ with $h_{1,0}(x) = v \in \mathbb{C}^2$. Thus in the regular case, $(f_0^* J h_0)(a) = f_0^*(a) J h_0(a) = 0$ for all $(h,k) \in \mathcal{T}$ is equivalent to $f_0(a) = 0$, as claimed. □

2.2 Von Neumann theory of symmetric relations

We now pause our analysis of the relations associated with a canonical system to develop the general theory of self-adjoint extensions of symmetric relations. For a general relation $\mathcal{T}$ and $z \in \mathbb{C}$, we define $\mathcal{N}_z = \{(f,zf) \in \mathcal{T}\}$, and if we want to emphasize which relation is being used, then we can also write $\mathcal{N}_z(\mathcal{T})$ for this. The fs that occur here can also be viewed as the null space $N(\mathcal{T} - z)$, if we interpret $z = zI = \{(f,zf) : f \in \mathcal{H}\}$ and define the sum of relations $\mathcal{S}$, $\mathcal{T}$ as

$$\mathcal{S} + \mathcal{T} = \{(f,g) : (f,h) \in \mathcal{S}, (f,g-h) \in \mathcal{T} \text{ for some } h\}.$$

This last definition is the obvious one if we again use the corresponding notion for operators as a guideline, but perhaps it is worth pointing out the (trivial) fact that this is of course not the same as taking the sum of the subspaces $\mathcal{S}$, $\mathcal{T}$.

Theorem 2.10 (Von Neumann formula). *Let* $\mathcal{T}_0$ *be a closed symmetric relation and write* $\mathcal{T} = \mathcal{T}_0^*$. *Then*

$$\mathcal{T} = \mathcal{T}_0 \oplus \mathcal{N}_i(\mathcal{T}) \oplus \mathcal{N}_{-i}(\mathcal{T}).$$

Proof. It is straightforward to check that these subspaces are orthogonal. For example, if $(f,g) \in \mathcal{T}_0$ and $(h,ih) \in \mathcal{N}_i$, then

$$\langle (f,g),(h,ih)\rangle = \langle f,h\rangle + \langle g,ih\rangle = -i(\langle f,ih\rangle - \langle g,h\rangle) = 0.$$

I leave the other two cases to the reader.

Clearly, the sum on the right-hand side is contained in $\mathcal{T}$. It remains to show the reverse inclusion. So let $(f,g) \in \mathcal{T}$ and suppose that $(f,g) \perp \mathcal{T}_0 \oplus \mathcal{N}_i$. We will show that then $(f,g) \in \mathcal{N}_{-i}$, and this will finish the proof since all three relations on the right-hand side are closed.

For any $(h,k) \in \mathcal{T}_0$, we have $\langle f,h\rangle = -\langle g,k\rangle$, from the orthogonality, but also $\langle f,k\rangle = \langle g,h\rangle$, since $\mathcal{T}_0$, $\mathcal{T}$ are adjoints of one another. It follows that $\langle f-ig,h\rangle = \langle -g-if,k\rangle$, and since $(h,k) \in \mathcal{T}_0$ is still arbitrary, this says that $(g+if, ig-f) \in \mathcal{T}$. But this element is of the form (p, ip), so in fact lies in $\mathcal{N}_i$. Hence it is orthogonal to (f,g), and by working out the corresponding scalar product, we find

$$i\|f\|^2 + \langle f,g\rangle + i\|g\|^2 - \langle g,f\rangle = i\|f-ig\|^2 = 0,$$

so $(f,g) = (f,-if) \in \mathcal{N}_{-i}$, as desired. □

Definition 2.4. The *deficiency indices* of the closed symmetric relation $\mathcal{T}_0$ with adjoint $\mathcal{T} = \mathcal{T}_0^*$ are defined as $\gamma_\pm = \dim \mathcal{N}_{\pm i}(\mathcal{T})$. Equivalently, $\gamma_\pm = \dim N(\mathcal{T} \mp i)$, and these latter spaces are also referred to as the *deficiency spaces*.

This terminology alludes to the fact that the spaces $\mathcal{N}_{\pm i}$ are what currently prevents our relation from being self-adjoint, if they are indeed present: we see from the von Neumann formula that $\mathcal{T}_0$ itself is self-adjoint precisely if $\gamma_+ = \gamma_- = 0$, by Theorem 2.10.

In general, we would now like to find the self-adjoint restrictions of $\mathcal{T}$, if any. If we restrict $\mathcal{T}$, then we at the same time extend its adjoint, so if there are self-adjoint realizations in this sense, then they will have to lie somewhere in the middle between $\mathcal{T}_0$ and $\mathcal{T}$. We will obtain a precise description in a moment, but let us first make sure that finding self-adjoint relations is actually a worthwhile goal. Especially in our original setting, for canonical systems, one is perhaps not so excited about self-adjoint relations and would much rather try to produce self-adjoint operators. Fortunately, there is not much of a difference between the two.

Theorem 2.11. *Let $\mathcal{T}$ be a self-adjoint relation and write $\mathcal{H}_1 = \overline{D(\mathcal{T})}$, $\mathcal{H}_2 = \mathcal{T}(0)$. Then $\mathcal{H} = \mathcal{H}_1 \oplus \mathcal{H}_2$, and this decomposition reduces $\mathcal{T}$ in the sense that*

$$\mathcal{T} = \mathcal{T}_1 \oplus \mathcal{T}_2, \quad \mathcal{T}_j := \mathcal{T} \cap (\mathcal{H}_j \oplus \mathcal{H}_j).$$

Moreover, $\mathcal{T}_1$ is a self-adjoint operator in $\mathcal{H}_1$, and $\mathcal{T}_2 = \{(0,g) : g \in \mathcal{H}_2\}$.

So a self-adjoint relation is a self-adjoint operator where it is defined, and then the orthogonal complement is tacked on as the multi-valued part. Or, put differently, a self-adjoint relation contains a self-adjoint operator on a possibly smaller space, which we obtain from the relation by simply dividing out the multi-valued part.

The decomposition of $\mathcal{H}$ is a special case of an easy slightly more general fact, which I will state separately.

Lemma 2.12. *For any relation $\mathcal{T}$, we have $D(\mathcal{T})^\perp = \mathcal{T}^*(0)$.*

Proof. We have $f \in D(\mathcal{T})^\perp$ if and only if $\langle f,h\rangle = 0$ for all $(h,k) \in \mathcal{T}$, but this is also the condition for $(0,f)$ to lie in $\mathcal{T}^*$. □

Proof of Theorem 2.11. By the lemma, $\overline{D(\mathcal{T})} = D(\mathcal{T})^{\perp\perp} = \mathcal{T}(0)^{\perp}$, so we do have the asserted decomposition of $\mathcal{H}$. Next, to show that $\mathcal{T} = \mathcal{T}_1 \oplus \mathcal{T}_2$, we must verify that every $(f, g) \in \mathcal{T}$ can be decomposed in this way, but by the nature of the multi-valued part $\mathcal{T}_2$, we really only need to check that for any $f \in D(\mathcal{T})$, there is an $(f, g) \in \mathcal{T}$ with $g \in \mathcal{H}_1$ also. This is obvious because we can just start out with an arbitrary $(f, g_0) \in \mathcal{T}$ and then add on a suitable element $(0, g) \in \mathcal{T}$ to make $g + g_0 \perp \mathcal{T}(0)$.

So we now know that the decomposition reduces $\mathcal{T}$, and this implies that the individual parts are self-adjoint separately. Finally, it is also clear that $\mathcal{T}_1$ is an operator; we are in the orthogonal complement of the original multi-valued part, so none of this remains. □

Definition 2.5. The *Cayley transform* $\mathcal{C} = \mathcal{C}_{\mathcal{T}}$ of a closed symmetric relation $\mathcal{T}$ is defined as

$$\mathcal{C} = \{(g + if, g - if) : (f, g) \in \mathcal{T}\}.$$

This is a key tool to study self-adjoint extensions of symmetric relations in an abstract setting. The definition we gave generalizes the (operator) Cayley transform $C = (T - i)(T + i)^{-1}$ of symmetric operators T to relations.

Proposition 2.13. *Let $\mathcal{T}_0$ be a closed symmetric relation, let $\mathcal{C}$ be its Cayley transform, and write $\mathcal{T} = \mathcal{T}_0^*$. Then $\mathcal{C}(0) = 0$, $N(\mathcal{C} - 1) = \mathcal{T}_0(0)$, and*

$$D(\mathcal{C})^{\perp} = N(\mathcal{T} - i) = R(\mathcal{T}_0 + i)^{\perp},$$
$$R(\mathcal{C})^{\perp} = N(\mathcal{T} + i) = R(\mathcal{T}_0 - i)^{\perp}.$$

Here, the relation sums $\mathcal{C} - 1$, $\mathcal{T} - i$, etc. must be interpreted as explained at the beginning of this section (generalize operator sums in the obvious way).

Proof. An $h \in \mathcal{C}(0)$ would have to come from an $(f, g) \in \mathcal{T}_0$ with $g = -if$, but this is not possible for a symmetric relation unless $f = g = 0$: this follows because for any $(f, g) \in \mathcal{T}_0$, the symmetry of $\mathcal{T}_0$ implies that $\langle f, g\rangle = \langle g, f\rangle$.

Next, $g \in N(\mathcal{C} - 1)$ means that $(g, g) \in \mathcal{C}$, which is indeed equivalent to $(0, g) \in \mathcal{T}_0$.

It is clear from the definition of $\mathcal{C}$ that $D(\mathcal{C}) = R(\mathcal{T}_0 + i)$, $R(\mathcal{C}) = R(\mathcal{T}_0 - i)$, and then we finish the proof by referring to the easy general fact stated in Lemma 2.14 below. □

Lemma 2.14. *For any relation $\mathcal{T}$, we have $N(\mathcal{T}^*) = R(\mathcal{T})^{\perp}$.*

Proof. The condition that $f \in N(\mathcal{T}^*)$ (or $(f, 0) \in \mathcal{T}^*$) is equivalent to $\langle f, k\rangle = 0$ for all $(h, k) \in \mathcal{T}$, and this says that $f \in R(\mathcal{T})^{\perp}$, as claimed. □

Theorem 2.15. *Let $\mathcal{T}$ be a closed symmetric relation. Then its Cayley transform $\mathcal{C}$ is isometric, that is, $\|h\| = \|k\|$ for all $(h, k) \in \mathcal{C}$. Moreover, $\mathcal{C}$ is closed, and so are the subspaces $D(\mathcal{C})$, $R(\mathcal{C})$ of $\mathcal{H}$.*

Clearly, an isometric relation cannot have a multi-valued part, so is an operator. For the Cayley transform, we knew this already from the first part of Proposition 2.13.

Proof. To prove that $\mathcal{C}$ is isometric, we must show that $\|g + if\|^2 = \|g - if\|^2$ for all $(f,g) \in \mathcal{T}$, but this is immediate from the fact that $\langle f,g\rangle = \langle g,f\rangle$, which we observed in the proof of Proposition 2.13. It is also fairly obvious that $\mathcal{C}$ is closed: if $(g_n + if_n, g_n - if_n) \to (h,k)$, with $(f_n, g_n) \in \mathcal{T}$, then f_n, g_n must also converge, so we obtain the desired conclusion from our assumption that $\mathcal{T}$ is closed. Finally, a closed isometric operator must have closed domain and range. □

We now have a clear picture of what the Cayley transform of a closed symmetric relation looks like: it is an operator and, more precisely, a partially defined isometry that maps the orthogonal complements of the deficiency spaces unitarily onto each other.

We now show that these properties characterize Cayley transforms, and one can go back from the Cayley transform to the original symmetric relation in just the same way as one would do this for operators.

Theorem 2.16. *Assume that the relation $\mathcal{C}$ is a closed isometry. Then $\mathcal{T} := \{(c-d, i(c+d)) : (c,d) \in \mathcal{C}\}$ is a closed symmetric relation with Cayley transform $\mathcal{C}_\mathcal{T} = \mathcal{C}$, and it is the only relation with these properties.*

Do not get confused by the terminology here: for operators, the term *isometry* would normally refer to an everywhere defined operator that preserves norms (and then also scalar products, by polarization), and such an operator is automatically closed. Here, for relations, an isometry just preserves norms and scalar products; it then is an operator, but it could be a partially defined one, with non-closed domain, and then it would not be closed.

Proof. The assumption that $\mathcal{C}$ is closed implies that $\mathcal{T}$ is closed also, in the same way as above, in the proof of Theorem 2.15. To verify that $\mathcal{T}$ is symmetric, we must show that

$$\langle a-b, i(c+d)\rangle = \langle i(a+b), c-d\rangle \quad \text{for all } (a,b),(c,d) \in \mathcal{C},$$

and this follows because $\langle a,c\rangle = \langle b,d\rangle$, by polarization. A simple calculation shows that taking the Cayley transform of $\mathcal{T}$ will get us back to $\mathcal{C}$, and the uniqueness claim is also clear, from the definition of the Cayley transform: if $\mathcal{C}_\mathcal{T}$ is a Cayley transform and $(c,d) \in \mathcal{C}_\mathcal{T}$, then it must be the case that $c = g + if$, $d = g - if$ for some $(f,g) \in \mathcal{T}$, but this uniquely determines (f,g) as $f = -i/2(c-d)$, $g = 1/2(c+d)$, so the $\mathcal{T}$ described in the theorem is indeed the only way to obtain $\mathcal{C}$ as a Cayley transform. □

Theorem 2.17. *Let $\mathcal{T}_j, \mathcal{T}$ be closed symmetric relations. Then the Cayley transforms satisfy $\mathcal{C}_1 \subseteq \mathcal{C}_2 \iff \mathcal{T}_1 \subseteq \mathcal{T}_2$, and $\mathcal{T}$ is self-adjoint if and only if $\mathcal{C}$ is unitary.*

Proof. The first claim is obvious from the definition of the Cayley transform in one direction and Theorem 2.16 in the other. By Theorem 2.10, $\mathcal{T}$ is self-adjoint if and only if both deficiency spaces are zero, and by Proposition 2.13 and Theorem 2.15, this happens if and only if $\mathcal{C}$ is unitary. □

These results now outline a clear-cut procedure how to find all self-adjoint realizations of a given closed symmetric relation $\mathcal{T}$. An equivalent assignment is to find all unitary extensions of the Cayley transform. This is the same as providing unitary maps from the first deficiency space onto the second. Let me summarize.

Corollary 2.18. *The closed symmetric relation $\mathcal{T}_0$ has self-adjoint extensions if and only if $\gamma_+ = \gamma_-$.*

Theorem 2.19. *Assume that the closed symmetric relation $\mathcal{T}_0$ has equal deficiency indices, and write $\mathcal{T} = \mathcal{T}_0^*$. Then for each unitary map $U : N(\mathcal{T} - i) \to N(\mathcal{T} + i)$, we obtain a self-adjoint extension $\mathcal{S} = \mathcal{S}_U$ of $\mathcal{T}_0$ which is given by*

$$\mathcal{S} = \mathcal{T}_0 \oplus \{(f - Uf, i(f + Uf)) : f \in N(\mathcal{T} - i)\}.$$

These are all self-adjoint extensions of $\mathcal{T}_0$.

Proof. These facts were already observed above, except perhaps for the formula for $\mathcal{S}$, which follows by going back from the Cayley transform to the symmetric (here: self-adjoint) relation, as described in Theorem 2.16. □

When we return to our analysis of self-adjoint realizations of canonical systems, we will actually not use this explicit description in terms of unitary maps between the deficiency spaces, but rather only the following general consequence: to produce the self-adjoint extensions, it suffices to get the dimension count right while keeping the relation symmetric. Let me state this more explicitly.

Theorem 2.20. *Let $\mathcal{T}_0$ be a closed symmetric relation with equal finite deficiency indices $\gamma_+ = \gamma_- = \gamma$. Then the self-adjoint extensions of $\mathcal{T}_0$ are exactly the γ-dimensional symmetric extensions of $\mathcal{T}_0$, or, equivalently, the γ-dimensional symmetric restrictions of $\mathcal{T}$.*

Here, *dimension* refers to the quotient spaces $\mathcal{S}/\mathcal{T}_0$, etc., or, since we have Hilbert spaces, we could take the (relative) orthogonal complements such as $\mathcal{S} \ominus \mathcal{T}_0$. Note also that by Theorem 2.10, $\dim \mathcal{T} \ominus \mathcal{T}_0 = 2\gamma$, so $\mathcal{T}$ is a 2γ-dimensional extension of $\mathcal{T}_0$ in this sense.

Proof. First of all, recall that a *finite-dimensional* extension or restriction of a closed relation is closed itself.

Next, we note that a self-adjoint realization $\mathcal{S}$ as described in Theorem 2.19 has the stated properties, by the von Neumann formula (Theorem 2.10) again. This formula also shows that a γ-dimensional extension of $\mathcal{T}_0$ (which is contained in $\mathcal{T}$, as it will be if the extension is symmetric) is the same as a γ-dimensional restriction of $\mathcal{T}$.

So we only need to show that a γ-dimensional symmetric extension $\mathcal{S}$ of $\mathcal{T}_0$ is self-adjoint. Since, as we saw in Theorem 2.16, the Cayley transform determines the symmetric relation, each extension by one dimension must also lead to a genuine extension of the Cayley transform, and the extension will have a strictly larger domain and range since we are dealing with isometric relations. But the original Cayley transform of $\mathcal{T}_0$ was already defined on a closed subspace whose orthogonal complement had dimension γ, so after this many one-dimensional such extensions we run out of space and thus the Cayley transform of $\mathcal{S}$ must be defined everywhere. It will then also be surjective, for similar reasons, and thus unitary, but this means that $\mathcal{S}$ itself is self-adjoint, as required. □

2.3 Self-adjoint realizations of canonical systems

We now apply this theory to the relations of a canonical system, as introduced in Section 2.1. Let us first identify the deficiency spaces. The condition that $(f, if) \in \mathcal{N}_i$ implies that $Jf_0' = -iHf_0$, so the elements of the deficiency space come from solutions of the differential equation $Ju' = -iHu$ that lie in $L^2_H(a,b)$. In fact, that is already the complete answer. To see this, notice that (unlike for solutions to inhomogeneous equations) if such a solution u represents the zero element of $L^2_H(a,b)$, then $u(x) = 0$ for all $x \in (a,b)$. Indeed, such a u would have to be constant, but a non-zero constant could only represent zero in the Hilbert space if (a,b) were a single singular interval. We can then also say that if L^2_H solutions of $Ju' = -iHu$ are linearly independent as functions, then they will also be linearly independent as elements of L^2_H.

These remarks imply that $\mathcal{T}_0$ has equal deficiency indices because we can take the complex conjugate to transform the solutions of the two equations $Ju' = \mp iHu$ into each other. We have established the following.

Proposition 2.21. *The minimal relation $\mathcal{T}_0$ has equal deficiency indices $\gamma_+ = \gamma_- = \gamma$, equal to the number of linearly independent $L^2_H(a,b)$ solutions of $Ju' = -iHu$.*

Since the whole solution space is only two-dimensional, the possible deficiency indices are 0, 1, and 2, and we will see later that all three values occur.

Theorem 2.22. *Suppose that $(f_j, g_j) \in \mathcal{T}$, $j = 1, \ldots, \gamma$, are linearly independent modulo $\mathcal{T}_0$ and satisfy $f_{j,0}^* J f_{k,0}|_a^b = 0$. Then*

$$\mathcal{S} = \{(h,k) \in \mathcal{T} : f_{j,0}^* J h_0|_a^b = 0,\ j = 1, \ldots, \gamma\} \tag{2.5}$$

is self-adjoint. Conversely, every self-adjoint realization is obtained in this way.

Proof. This basically spells out what Theorem 2.20 has to say in the case at hand. Clearly, $(f_j, g_j) \in \mathcal{S}$ for all j, so $\mathcal{S}$ is an at least γ-dimensional extension of $\mathcal{T}_0$. On the other hand, it is also an at least γ-dimensional restriction of $\mathcal{T}$: this follows because the functionals $F_j : \mathcal{T} \to \mathbb{C}$ that we use to define $\mathcal{S}$ as $\mathcal{S} = \mathcal{T} \bigcap N(F_j)$ are linearly independent, as we see from Theorem 2.9. So $\mathcal{S}$ is an exactly γ-dimensional extension of $\mathcal{T}_0$. This argument has also shown that the (f_j, g_j) together with $\mathcal{T}_0$ span $\mathcal{S}$, so $\mathcal{S}$ is symmetric (use Theorem 2.9 again) and thus self-adjoint by Theorem 2.20.

Conversely, if a self-adjoint (and thus γ-dimensional) extension $\mathcal{S}$ of $\mathcal{T}_0$ is given, then just pick γ elements $(f_j, g_j) \in \mathcal{S}$ that are linearly independent modulo $\mathcal{T}_0$. These then will satisfy $f_{j,0}^* J f_{k,0}|_a^b = 0$, since $\mathcal{S}$ is symmetric, and we can use them to define a relation $\mathcal{S}_1$ by (2.5). Clearly, $\mathcal{S} \subseteq \mathcal{S}_1$, and $\mathcal{S}_1$ is also self-adjoint, by the first part of the proof, so taking adjoints shows that $\mathcal{S} = \mathcal{S}_1$. □

Since the value of $f_0^* J h_0$ at an endpoint, for a given f_0, only depends on the behavior of h_0 near that endpoint, we can think of the extra conditions of Theorem 2.22 that describe which elements of $\mathcal{T}$ are in $\mathcal{S}$ as *boundary conditions*. This works especially well at regular endpoints; then $f_0^* J h_0$ literally only depends on the value of (the unique continuous extension of) h_0 at that endpoint. Another general remark is that f_0 could be such that $f_0^* J h_0 = 0$ at one endpoint for all $(h, k) \in \mathcal{T}$, and then the corresponding boundary condition only involves the other endpoint. For example, this happens if $f_0 = 0$ near one endpoint. In fact, as we will see later, *coupled boundary conditions*, that is, boundary conditions that involve both endpoints, only occur if $\gamma = 2$.

We now prove an important result that says that each endpoint makes a separate contribution to the deficiency index, and there is no interaction between these. For an endpoint $t = a, b$, define γ_t as the number of linearly independent solutions of $Ju' = -iHu$ that are in L_H^2 in a neighborhood of t.

Theorem 2.23. *For each endpoint $t = a, b$, we have $\gamma_t = 1$ or $\gamma_t = 2$, and the deficiency index γ satisfies $\gamma = \gamma_a + \gamma_b - 2$.*

As I mentioned earlier, this is best interpreted as saying that each endpoint separately makes the contribution of $\gamma_t - 1$ to the deficiency index.

Proof. I will discuss explicitly only the case where it is possible to find a point $c \in (a, b)$ such that neither (a, c) nor (c, b) is contained in a single singular interval; this is the same as assuming that (a, b) does not consist of just two consecutive singular intervals. Fix such a c and consider the relation $\mathcal{T}_c = \mathcal{T}_{(a,c)} \oplus \mathcal{T}_{(c,b)}$ and its adjoint $\mathcal{T}_{c,0} = \mathcal{T}_c^*$. These can naturally be viewed as relations on $L_H^2(a, b)$ by referring to the decomposition $L_H^2(a, b) = L_H^2(a, c) \oplus L_H^2(c, b)$. Since c is a regular endpoint of both smaller intervals, this remark, when combined with Theorem 2.9 and Lemma 2.7, makes it clear that

$$\mathcal{T}_{c,0} = \{(f, g) \in \mathcal{T}_0 : f_0(c) = 0\}.$$

Now pick two elements $(f_j, g_j) \in \mathcal{T}_0$ with $f_{j,0}(c) = e_j$, $j = 1, 2$. We can easily construct such elements with the help of Lemma 2.7: for example, make $f_{j,0} = e_j$ near c and $f_{j,0} = 0$ near both endpoints. I now claim that

$$\mathcal{T}_0 = \mathcal{T}_{c,0} \dotplus L((f_1, g_1)) \dotplus L((f_2, g_2)). \tag{2.6}$$

Obviously, the sum on the right-hand side is contained in $\mathcal{T}_0$, and it is also clear that it is direct because a non-trivial linear combination $(f, g) = c_1(f_1, g_1) + c_2(f_2, g_2)$ will satisfy

$f_0(c) \neq 0$. Conversely, if $(f, g) \in \mathcal{T}_0$, then subtraction of a suitable linear combination of the (f_j, g_j) will produce an element (h, k) with $h_0(c) = 0$, so $(h, k) \in \mathcal{T}_{c,0}$.

We learn from (2.6) that $\mathcal{T}_0$ is a two-dimensional extension of $\mathcal{T}_{c,0}$, and this then implies the corresponding situation for the adjoints also: $\mathcal{T}_c$ is a two-dimensional extension of $\mathcal{T}$. Thus the deficiency indices satisfy $\gamma(\mathcal{T}_{c,0}) = \gamma(\mathcal{T}_0) + 2$. Since $\mathcal{T}_{c,0}$ is an orthogonal sum of two independent relations, we can compute its deficiency index as the sum of those of $\mathcal{T}_{(a,c),0}$ and $\mathcal{T}_{(c,b),0}$, but clearly these equal γ_a and γ_b, respectively, by Proposition 2.21 and since every solution of $Ju' = -iHu$ is in L^2_H near the regular endpoint c.

In the remaining case, when (a, b) consists of two singular intervals, this argument does not apply (the first and last steps are not valid), but in this situation, the statement is easy to verify directly, by distinguishing subcases according to whether or not $H \in L^1$ near the endpoints. I leave this to the reader.

To prove that it is not possible to have $\gamma_a = 0$ (or $\gamma_b = 0$), we use the (unsurprising) fact that there are canonical systems with $\gamma_t = 1$ at an endpoint, as we can confirm with simple concrete examples that can be worked out explicitly, such as $H = \left(\begin{smallmatrix} 1 & 0 \\ 0 & 0 \end{smallmatrix}\right)$ on $(0, \infty)$, say. We now take the $H(x)$ we had originally on (a, c) and then glue on such an example on (c, b'), so that γ_a is not affected by this modification, but $\gamma_{b'} = 1$ now. Then, if we had $\gamma_a = 0$, then what we just proved would give the absurd conclusion that the modified system has deficiency index $\gamma = 0 + 1 - 2 = -1$. □

Definition 2.6. We say that an endpoint t is in the *limit point case* if $\gamma_t = 1$, and in the *limit circle case* if $\gamma_t = 2$.

This terminology may sound strange in our current context (what points or circles? or limits, for that matter?), but of course there are reasons for it, and it will be justified in Chapter 3, when we discuss Weyl theory.

Theorem 2.23 then says that each limit circle endpoint (if any) makes a contribution of 1 to the deficiency index. Regular endpoints are always in the limit circle case because solutions stay bounded near such an endpoint and H is integrable there.

Theorem 2.24. *An endpoint $t = a, b$ is in the limit point case if and only if $(f_0^* J h_0)(t) = 0$ for all $(f, g), (h, k) \in \mathcal{T}$.*

Proof. We proceed as in the last part of the proof of Theorem 2.23. Let's say $t = a$. We define a new $H(x)$ by starting out with the one we had on (a, c), with $c \in (a, b)$, and then continuing with a new $H(x)$ that gives us limit point case at the new right endpoint b'. By Lemma 2.7, we obtain the same collection of functions when we restrict the f_0 from (f, g) to a neighborhood of a, whether we use the original maximal relation or the one of the modified system. Thus the statement of the theorem is not affected by our modification. If we now have limit point case at a, then the modified system has deficiency index 0, so the claim follows from Theorem 2.9. Conversely, assume that $(f_0^* J h_0)(a) = 0$ always. By the implication we just established, we also have

$(f_0^* J h_0)(b') = 0$, and now Theorem 2.9 implies that the minimal and maximal relations (of the modified system) agree, so $\gamma = 0$ and thus $\gamma_a = 1$ by Theorem 2.23. □

This result gives considerable additional insight into the nature of boundary conditions, as discussed after the statement of Theorem 2.22. If $\gamma = 1$, equivalently, if there is one limit point and one limit circle endpoint (and let us say this is a), then the single boundary condition $f_0^* J h_0|_a^b = 0$ becomes $(f_0^* J h_0)(a) = 0$, so only restricts our functions near the limit circle endpoint a. Coupled boundary conditions, that involve both endpoints, only occur if we have two limit circle endpoints.

Finally, let us work out the general form of a separated boundary condition at a regular endpoint. This will be easy to do, but before we actually do it, let us formulate the question in more precise terms. Let us make a the regular endpoint. We then consider a separated boundary condition, described as in Theorem 2.22 by an $(f, g) \in \mathcal{T}$ with $f_0^* J f_0|_a^b = 0$, and *separated* means that

$$(f_0^* J h_0)(b) = 0 \quad \text{for all } (h, k) \in \mathcal{T}, \tag{2.7}$$

so that the boundary condition is really only imposed at a. Since a is regular, it becomes $f_0^*(a) J h_0(a) = 0$. Theorem 2.22 also requires us to take $(f, g) \notin \mathcal{T}_0$, and for an (f, g) satisfying (2.7), this simply means that $f_0(a) \neq 0$. Since for any $v \in \mathbb{C}^2$, there are (f, g) as described with $f_0(a) = v$ (use Lemma 2.7 again to see this), we now have a precise version of our question, and it simply is: what $v \in \mathbb{C}^2$, $v \neq 0$, satisfy $v^* J v = 0$?

Answering the question is much easier than finding it was. A quick calculation shows that $v^* J v = 2i \operatorname{Im} v_1 \overline{v_2}$, and this equals zero precisely if $v = sw$, with $w \in \mathbb{R}^2$, $s \in \mathbb{C}$. The factor s is of course irrelevant as far as the boundary condition is concerned; we could use it to normalize v and make the entries real, and then the boundary conditions at a regular endpoint are exactly obtained from the vectors $v = e_\alpha = (\cos\alpha, \sin\alpha)^t$. The corresponding boundary condition becomes $e_\alpha^* J f_0(a) = 0$, and what it is asking for is that $f_0(a)$ be a multiple of e_α.

Let us apply this to two scenarios, with a reminder on a third important case thrown in for good measure. Right now, these look like special cases because we do not allow non-regular limit circle endpoints. However, that is not the case: we will later obtain the beautiful result that canonical systems cannot have limit circle endpoints that are not regular (Corollary 3.6). So, we are about to see the sum of our labors in this chapter: the description of the self-adjoint realizations in all cases.

Theorem 2.25. (a) *Suppose that both endpoints are limit point. Then $\gamma = 0$, and $\mathcal{T}_0 = \mathcal{T}$ is self-adjoint.*

(b) *Suppose that a is regular and b is limit point. Then $\gamma = 1$. For every $\alpha \in [0, \pi)$, the restriction of $\mathcal{T}$ by the boundary condition*

$$e_\alpha^* J f_0(a) = f_{0,1}(a) \sin\alpha - f_{0,2}(a) \cos\alpha = 0$$

defines a self-adjoint relation, and these are all self-adjoint realizations.

(c) *Suppose that both endpoints are regular. Then* $\gamma = 2$, *and the self-adjoint realizations with separated conditions are exactly those described by*

$$e_\alpha^* J f_0(a) = e_\beta^* J f_0(b) = 0,$$

with $\alpha, \beta \in [0, \pi)$. *There are other self-adjoint realizations with coupled boundary conditions.*

An example for such coupled boundary conditions, in case (c), would be $f_0(a) = f_0(b)$; note that since f_0 has two components, this is really two conditions. It is in principle straightforward to describe *all* boundary conditions in this case also: similarly to what we did above, one must now find all linearly independent pairs of vectors $V_j = (v_j, w_j)^t \in \mathbb{C}^4$, $j = 1, 2$, such that $V_j^* \mathcal{J} V_k = 0$, where $\mathcal{J}$ denotes the block matrix

$$\mathcal{J} = \begin{pmatrix} J & 0 \\ 0 & -J \end{pmatrix}.$$

We will not use coupled boundary conditions in this book, and I will leave the matter at that.

It is perhaps also instructive to now recall Theorem 2.19, which said that the self-adjoint realizations are parametrized by the unitary maps between the deficiency spaces. If $\gamma = 1$, then we are dealing with $U(1) \cong S^1$, and what we found by different methods in Theorem 2.25(b) fits well into this general picture: the self-adjoint realizations we found are indeed naturally parametrized by the unit circle. Similarly, in part (c), if it had not been clear right away, Theorem 2.19 also would have made it unlikely that these are all self-adjoint realizations since they are naturally parametrized by two circles (a torus), which seems a far cry from $U(2)$.

2.4 The multi-valued part

We now determine the multi-valued part of our relations. This can be done quite explicitly, and the quick summary is that this happens exclusively on the singular intervals, and each singular interval (c, d) contributes a codimension 1 subspace of $L_H^2(c, d)$ to $\mathcal{T}(0)$. Also, there are small modifications to these statements if the singular interval begins or ends at an endpoint. This does not introduce any real difficulties, but it does make things somewhat awkward to state.

Before we get started, let me introduce a piece of notation that will be used throughout in the sequel. Namely, for an angle α, we will write

$$e_\alpha = \begin{pmatrix} \cos\alpha \\ \sin\alpha \end{pmatrix}, \quad e_{\alpha^\perp} = \begin{pmatrix} -\sin\alpha \\ \cos\alpha \end{pmatrix}.$$

The first of these notations has been used before on occasion; the sign of the orthogonal direction has been chosen so that $e_{\alpha^\perp} = J e_\alpha$, though that is actually not essential.

Now let us return to our discussion of the multi-valued part. We begin with the following key observation.

Lemma 2.26. *Let $(0,g) \in \mathcal{T}$. Then $f_0(x) = 0$ at all regular points $x \in (a,b)$.*

Here, f_0 as usual refers to the unique representative of $f = 0 \in L^2_H$ determined by $Jf_0' = -Hg$. It would have been even more consistent to denote this by 0_0, but this looks too strange.

Proof. Let $x_0 \in (a,b)$ be a point with $f_0(x_0) \neq 0$. We must show that such an x_0 is singular. Since f_0 is (absolutely) continuous, we have $f_0 \neq 0$ in a neighborhood of x_0, so we can write $f_0(x) = R(x)e_{\varphi(x)}$ there, with $R(x) > 0$, $\varphi(x) \in \mathbb{R}$, and with φ continuous. Then R, φ are in fact absolutely continuous themselves, and

$$f_0' = R'e_\varphi + R\varphi' e_{\varphi^\perp}. \tag{2.8}$$

Since f_0 represents the zero element but $R > 0$ throughout our neighborhood, $H(x)$ must be singular and in fact $H(x) = h(x)P_{\varphi(x)^\perp}$ there. So if we plug (2.8) into $f_0' = JHg$ and take the scalar product with $e_{\varphi^\perp}$, we find $R\varphi' = 0$, so φ is constant near x_0, and that makes x_0 a singular point. □

Theorem 2.27. *Consider a canonical system with singular intervals (c_j, d_j) and write $H(x) = h_j(x)P_{\alpha_j}$ on $x \in (c_j, d_j)$.*

Let $g \in L^2_H(a,b)$. Then $g \in \mathcal{T}(0)$ if and only if

$$g(x) = \begin{cases} r_j(x)e_{\alpha_j}, & x \in (c_j, d_j), \\ 0, & x \text{ regular}, \end{cases}$$

with $\int_{c_j}^{d_j} h_j(x)r_j(x)\,dx = 0$ for all those singular intervals with $c_j \neq a$, $d_j \neq b$.

A more explicit version of the statement would be to say that $g \in L^2_H$ lies in $\mathcal{T}(0)$ if and only if it is represented by a function of the form given. Also, recall that we refer to α_j as the type of the singular interval (c_j, d_j); see Definition 1.2.

Proof. Suppose first that $g \in \mathcal{T}(0)$. Then we want to deduce from Lemma 2.26 that $Hg = -Jf_0' = 0$ at regular points, but some care is required here since these might be surrounded by singular points to which the Lemma does not apply. We argue as follows: if we denote the set of regular points by R, then almost every $x \in R$ is a point of density of R, so the classical derivative $f_0'(x)$ at such a point may be computed by touching regular points only, if it exists at all (but it will at almost all x). Thus $f_0'(x) = 0$ under these assumptions, and we do obtain the desired conclusion.

It remains to show that g also has the stated properties on a singular interval (c,d). Since $H = hP_\alpha$ there, any $g \in L^2_H(c,d)$ can be represented by a function of the form $r(x)e_\alpha$, and we only need to verify that $\int_c^d hr\,dx = 0$. This follows because if $c \neq a$, $d \neq b$, then c, d are regular points, so the f_0 from $Jf_0' = -Hg$ must satisfy $f_0(c) = f_0(d) = 0$. Since

$$f_0(x) = f_0(c) + J\int_c^x H(t)g(t)\,dt,$$

this leads to the desired conclusion.

Conversely, if g is of the form given and $g \in L_H^2(a,b)$, we simply pick a regular $c \in (a,b)$ and define, as we must, a function $f(x)$ by

$$f(x) = J\int_c^x H(t)g(t)\,dt. \tag{2.9}$$

Recall first of all that $Hg \in L^1_{\text{loc}}$, so this is well defined. The integral over any complete singular interval vanishes, by assumption, so $f(x) = 0$ at all regular $x \in (a,b)$. As a consequence, for any singular interval (c_j, d_j), we may change c to c_j in (2.9), and the formula still works. Thus $f(x)$ is a (non-constant) multiple of $e_{\alpha_j^\perp}$ on (c_j, d_j), so $Hf = 0$ there, too. (If $c_j = a$, then this argument still works, but then we of course take $c = d_j$ in (2.9).)

So the f from (2.9) represents the zero element of $L_H^2(a,b)$, and, by construction, it satisfies $Jf' = -Hg$. Thus $g \in \mathcal{T}(0)$, as claimed. □

While this explicitly talks only about the maximal relation, we of course obtain the multi-valued part of all other relations in the same way. Let us look at the case where (a,b) starts with a singular interval (a,c) and a is regular. In this situation, I want to work out $\mathcal{T}_\alpha(0)$, with $\mathcal{T}_\alpha$ defined as the one-dimensional restriction of $\mathcal{T}$ by the boundary condition α at a. If b is in the limit point case, then these are the self-adjoint realizations $\mathcal{S}_\alpha$; if b is in the limit circle case, then the self-adjoint restrictions of $\mathcal{T}_\alpha$ are exactly the self-adjoint realizations of the original relation with separated boundary conditions.

Before we do this, it is also useful to observe that, in general, if (a,b) neither starts nor ends with a singular interval, then nothing changes and $\mathcal{T}_0(0) = \mathcal{T}(0)$. To see this, recall that exactly those $(f,g) \in \mathcal{T}$ with $f_0^* J h_0 = 0$ at both endpoints for all $(h,k) \in \mathcal{T}$ are in $\mathcal{T}_0$. This will always be true for $f = 0$ because then $f_0 = 0$ at all regular points by Lemma 2.26, and there are regular points arbitrarily close to either endpoint.

Now assume that (a,c) is a singular interval for some $c \in (a,b)$, so $H = h(x)P_\beta$ there. We showed in the proof of Theorem 2.27 that if $(0,g) \in \mathcal{T}$ and $g = r(x)e_\beta$ on (a,c), then

$$f_0(x) = -\left(\int_x^c h(t)r(t)\,dt\right)e_{\beta^\perp}, \quad x \in (a,c).$$

We must now distinguish two cases. (1) Suppose that $\alpha \neq \beta^\perp$, with α still denoting the boundary condition at a. Then $e_{\beta^\perp}$ does not satisfy the boundary condition at a. So g will lie in $\mathcal{T}_\alpha(0)$ only if the additional condition $\int_a^c hr\,dx = 0$ is satisfied.

This identifies the multi-valued part in this case, and we have found that $\dim \mathcal{T}(0) \ominus \mathcal{T}_\alpha(0) = 1$, and a possible choice for the extra vector in $\mathcal{T}(0)$ would be $g(x) = \chi_{(a,c)} e_\beta$. The argument has also shown that if $(0,g) \in \mathcal{T}$ and f_0 satisfies the boundary condition at $x = a$, then in fact $f_0(a) = 0$. So if we denote the restriction of $\mathcal{T}$ by this condition $f_0(a) = 0$ by $\mathcal{T}_0^{(a)}$, then $\mathcal{T}_\alpha(0) = \mathcal{T}_0^{(a)}(0)$.

(2) On the other hand, if $\alpha = \beta^\perp$, then the boundary condition at a is satisfied no matter what we do, so $\mathcal{T}_\alpha(0) = \mathcal{T}(0)$ in this case.

Let us summarize.

Theorem 2.28. (a) *If (a,b) neither starts nor ends with a singular interval, then $\mathcal{T}_0(0) = \mathcal{T}(0)$, and thus also $\mathcal{S}(0) = \mathcal{T}(0)$ for all self-adjoint realizations $\mathcal{S}$.*

(b) *Suppose that a is regular and (a,b) starts with a singular interval (a,c) of type β. Then*

$$\mathcal{T}(0) = \mathcal{T}_0^{(a)}(0) \oplus L(\chi_{(a,c)} e_\beta).$$

Moreover, the boundary condition α at $x = a$ will have the following effect:

(i) *if $\alpha = \beta^\perp$, then $\mathcal{T}_\alpha(0) = \mathcal{T}(0)$;*

(ii) *if $\alpha \neq \beta^\perp$, then $\mathcal{T}_\alpha(0) = \mathcal{T}_0^{(a)}(0)$.*

In particular, in the situation of part (b), exactly those $g \in \mathcal{T}(0)$ with $\int_a^c hr = 0$ are in $\mathcal{T}_0^{(a)}(0)$; in other words, the condition from Theorem 2.27 is now imposed on the interval (a,c) also.

The scenario not covered by this, when (a,b) starts with a singular interval and also ends with one, is handled in the same way, by applying the method behind case (b) to both endpoints in two separate steps. The reader might also wonder about limit point endpoints that bound an initial (or final) singular interval, but then there is no difference between $D(\mathcal{T}_0)$ and $D(\mathcal{T})$ as far the behavior of our functions near that endpoint is concerned, by Theorem 2.24, and thus the issue disappears.

We have here also seen an instance of a general phenomenon: if our interval (a,b) at a regular endpoint starts with a singular interval and the boundary condition e_α we impose there is exactly the direction annihilated by H, then the whole singular interval has no effect. It is easy in fact to prove a precise general statement along these lines.

Theorem 2.29. *Suppose that a is a regular endpoint and (a,c) is a singular interval of type $\alpha^\perp$. Impose boundary condition α at a, and also a boundary condition at b if b is limit circle, to obtain a self-adjoint realization $\mathcal{S}$. Then*

$$\mathcal{S} = (0, L^2_H(a,c)) \oplus \mathcal{S}_c,$$

where $\mathcal{S}_c$ denotes the self-adjoint realization on (c,b) with the same boundary condition α, but now at c (and also the same boundary condition at b as before, if any).

So essentially such a singular interval just pushes back the boundary condition to c. If we extract a self-adjoint operator from the relation $\mathcal{S}$, using the recipe from Theorem 2.11, then (a,c) makes no contribution to the Hilbert space $\mathcal{H}_1$ our operator will act in, and we will have $\mathcal{H}_1 \subseteq L^2_H(c,b)$. One might be tempted to just remove such an interval, but we will not do this since it does have an effect on the m function, which will become one of our primary objects.

Proof. This follows from the same calculations as above. If $(f,g) \in \mathcal{T}$, then $f_0(x)-f_0(c)$ is a multiple of e_α on (a,c), so f_0 will satisfy the boundary condition α at a precisely if it satisfies it at c. Moreover, this condition is also equivalent to f_0 representing the zero element of $L^2_H(a,c)$. It remains to confirm to $(0,g) \in \mathcal{S}$ for all $g \in L^2_H(a,c)$, which we do in the usual way by just defining a representative of zero by

$$f(x) = -J\int_x^c H(t)g(t)\,dt,$$

and then $f(x) = 0$ for $x > c$. This does satisfy the boundary condition at a, as we just discussed, it represents the zero element of $L^2_H(a,b)$, and $Jf' = -Hg$. Hence $g \in \mathcal{S}(0)$, as claimed. □

2.5 More on deficiency indices

We now return to the general theory and consider an arbitrary closed symmetric relation $\mathcal{T}_0$, with adjoint $\mathcal{T} = \mathcal{T}_0^*$. In the definition of the deficiency spaces $N(\mathcal{T} \mp i)$, we can replace i by a general number $z \in \mathbb{C}$. This extra generality is not needed to study self-adjoint realizations, as we saw, but the results of this section will come in handy in the next chapter. They are also of some interest in their own right.

Definition 2.7. Let $\mathcal{T}_0$ be a closed symmetric relation and write $\mathcal{T} = \mathcal{T}_0^*$. The *regularity domain* of $\mathcal{T}_0$ is the set

$$\Gamma(\mathcal{T}_0) = \{z \in \mathbb{C} : \text{ there is a } \delta = \delta(z) > 0 \text{ such that } \|g - zf\| \geq \delta\|f\| \text{ for all } (f,g) \in \mathcal{T}_0\}.$$

For $z \in \mathbb{C}$, we define

$$\gamma(z) = \dim N(\mathcal{T} - z) = \dim R(\mathcal{T}_0 + \overline{z})^\perp.$$

The last equality follows from Lemma 2.12. The definition of $\gamma(z)$ generalizes our earlier definition of the deficiency indices $\gamma_\pm$ in Definition 2.4; in the new notation, $\gamma_\pm = \gamma(\pm i)$.

Proposition 2.30. *The regularity domain $\Gamma(\mathcal{T}_0)$ is an open set and $\Gamma(\mathcal{T}_0) \supseteq \mathbb{C} \setminus \mathbb{R}$.*

Proof. If $z_0 \in \Gamma(\mathcal{T}_0)$ and $z \in \mathbb{C}$ is arbitrary, then

$$\|g - zf\| \geq \|g - z_0 f\| - |z - z_0|\|f\| \geq (\delta - |z - z_0|)\|f\|,$$

so $z \in \Gamma(\mathcal{T}_0)$ as well if $|z - z_0| < \delta$.

Let $z = x + iy \in \mathbb{C} \setminus \mathbb{R}$. We observed earlier that $\langle f, g\rangle = \langle g, f\rangle$ for all (f, g) from a symmetric relation, and now a quick calculation shows that

$$\|g - zf\|^2 = \|g - xf\|^2 + y^2\|f\|^2$$

for $(f, g) \in \mathcal{T}_0$, so $\delta = |y|$ works, and $z \in \Gamma(\mathcal{T}_0)$, as claimed. □

Theorem 2.31. *The function $\gamma(z)$ is constant on each connected component of $\Gamma(\mathcal{T}_0)$.*

In particular, $\gamma(z)$ is constant on each half plane $\mathbb{C}^{\pm} = \{z \in \mathbb{C} : \pm\text{Im}\, z > 0\}$. This also implies that if $\Gamma(\mathcal{T}_0)$ contains a real number, then $\mathcal{T}_0$ has equal deficiency indices (and thus self-adjoint extensions).

Proof. It suffices to show that $\gamma(z)$ is locally constant, that is, for each $z_0 \in \Gamma(\mathcal{T}_0)$, there exists a neighborhood of z_0 such that $\gamma(z) = \gamma(z_0)$ there. Indeed, this will establish that $\gamma : \Gamma(\mathcal{T}_0) \to \mathbb{N} \cup \{\infty\}$ is continuous, so will map a connected set to a connected set again.

Let $w, z \in \Gamma(\mathcal{T}_0)$. If $\gamma(w) < \gamma(z)$, then we can find an $x \in N(\mathcal{T} - z)$, $x \perp N(\mathcal{T} - w)$, $\|x\| = 1$. Equivalently, $x \in \overline{R(\mathcal{T}_0 - \overline{w})}$, $x \perp R(\mathcal{T}_0 - \overline{z})$. The first condition says that there is a sequence $(f_n, g_n) \in \mathcal{T}_0$ such that $y_n := g_n - \overline{w}f_n - x$ satisfies $y_n \to 0$. By the second condition, $\langle x, g_n - \overline{z}f_n\rangle = 0$, so

$$\begin{aligned} \langle y_n, g - \overline{z}f_n\rangle &= \langle g_n - \overline{w}f_n, g_n - \overline{z}f_n\rangle \\ &= \|g_n - \overline{w}f_n\|^2 + (\overline{w} - \overline{z})\langle g_n - \overline{w}f_n, f_n\rangle. \end{aligned} \tag{2.10}$$

Either $\overline{w} = w$, or $w \notin \mathbb{R}$. In either case, $\overline{w} \in \Gamma(\mathcal{T}_0)$, so $\|g_n - \overline{w}f_n\| \geq \delta\|f_n\|$ for some $\delta = \delta(w) > 0$. Since $\|g_n - \overline{w}f_n\| \to 1$, this shows that $\limsup |\langle g_n - \overline{w}f_n, f_n\rangle| \leq 1/\delta$. Thus (2.10) implies that if $|z - w| \leq \delta/3$, say, then

$$|\langle y_n, g_n - \overline{z}f_n\rangle| \geq \frac{1}{2}$$

for all large n. However, the left-hand side converges to zero, since $y_n \to 0$ and $g_n - \overline{z}f_n$ stays bounded, so this is impossible. The conclusion of all this is that if $|z-w| < \delta(w)/3$, then the situation assumed at the beginning of this paragraph will not occur, and thus $\gamma(z) \leq \gamma(w)$ for these z.

Since we can take $\delta(z) \geq \delta(w)/2$, say, for all z sufficiently close to w, it then also follows from this, by letting w and z switch roles, that $\gamma(z) = \gamma(w)$ on a (possibly smaller) neighborhood of w. □

2.6 Notes

The whole treatment of this chapter is modeled on the rather classical theory of self-adjoint realizations of differential expressions, except that in this classical case, one works with operators exclusively. A rather comprehensive discussion of the theory in this setting may be found in [65].

As discussed, relations are really needed here, for canonical systems, and they provide a flexible and easy to use tool. Therefore, I keep them all the way through, and in this respect my treatment differs slightly from what can be found in the literature. In fact, even though the material of this chapter would seem to form the foundation of everything else, it is not often discussed explicitly and in full detail. This is possible because one can just start out with the end results, so to speak, that is, impose boundary conditions where needed and reduce the Hilbert space by dividing out the multi-valued part (without calling it that), without further justification. In fact, since the central object of spectral theory is the Titchmarsh–Weyl m function, which can also be introduced by hand, without having defined self-adjoint relations before, one does not even need relations or operators at all to define the basic objects.

Such a pragmatic approach is often convenient in research papers, to keep the introduction and setting of the stage short, but of course it is not appropriate here. One also runs the risk of becoming confused about simple basic points if relations are not given their proper place. For example, boundary conditions for canonical systems are not imposed on elements $f \in D(\mathcal{T})$ of the domain (this would be plain nonsense as we do not have enough information to uniquely determine the representative f_0), but on pairs $(f, g) \in \mathcal{T}$.

An analysis of the relations of a canonical system similar to the one given in this chapter is contained in [29], and see also [33]; the general theory of self-adjoint extensions of symmetric relations is discussed in [21].

3 Spectral representation

3.1 Spectral representation of relations

We saw in Theorem 2.11 of the previous chapter that a self-adjoint relation $\mathcal{S}$ is really a self-adjoint operator on the possibly smaller Hilbert space $\mathcal{H}_1 = \overline{D(\mathcal{S})}$, plus $\mathcal{H}_1^\perp$ added on as the multi-valued part. For the spectral analysis of $\mathcal{S}$, it is now tempting to divide out $\mathcal{H}_1^\perp$, focus on the self-adjoint operator $S = \mathcal{S} \cap (\mathcal{H}_1 \oplus \mathcal{H}_1)$ on $\mathcal{H}_1$, and be done with relations once and for all. We will not do this here. As we will see, relations are a flexible tool that can handle the spectral theory quite well, and the extra generality gained is sometimes useful; for example, it may be convenient to work in the original Hilbert space $\mathcal{H} = L_H^2(a, b)$ without having to subtract the multi-valued part first.

For these reasons, we now generalize some fundamental spectral theoretic notions to the relations setting.

Definition 3.1. Let $\mathcal{S}$ be a self-adjoint relation. Then we define the *spectrum* $\sigma(\mathcal{S})$ as the spectrum $\sigma(S)$ of its operator part S if $D(\mathcal{S}) \neq 0$, and if $D(\mathcal{S}) = 0$, then we put $\sigma(\mathcal{S}) = \emptyset$. A *spectral representation* of $\mathcal{S}$ is a collection of Borel measures ρ_j on $\mathbb{R}$, together with a partial isometry $U : \mathcal{H} \to \bigoplus L^2(\mathbb{R}, \rho_j)$, with the following properties:

(1) $N(U) = \mathcal{S}(0)$;

(2) U maps $\overline{D(\mathcal{S})}$ unitarily onto $\bigoplus L^2(\mathbb{R}, \rho_j)$;

(3)
$$U\mathcal{S}U^* = M_t,$$
and here M_t denotes the (self-adjoint) operator of multiplication by the variable in $\bigoplus L^2(\mathbb{R}, \rho_j)$ on its natural domain.

The measures ρ_j are called *spectral measures* of S and $\mathcal{S}$, and the *spectral multiplicity* is defined as the minimal number of $L^2(\mathbb{R}, \rho_j)$ spaces in a spectral representation.

Several details here could have been handled differently; our version is tailor-made to fit the spectral representations we will actually construct later in this chapter. Note that the measures ρ_j are not required to be finite. The ones that we obtain below will satisfy $\int \frac{d\rho(t)}{1+t^2} < \infty$.

In the last (and crucial) condition, we compose the *relations* on the left-hand side, but the result is the *operator* M_t (and, as usual, we identify operators with their graphs without further comment). The product of relations is defined in the obvious way: first of all, we now work with relations between distinct Hilbert spaces $\mathcal{H}_1$, $\mathcal{H}_2$, which we of course define as linear subspaces of $\mathcal{H}_1 \oplus \mathcal{H}_2$. All basic definitions have obvious extensions to this new setting. Then the product $\mathcal{AB}$ of relations $\mathcal{A} \in \mathcal{H}_2 \oplus \mathcal{H}_3$, $\mathcal{B} \in \mathcal{H}_1 \oplus \mathcal{H}_2$ is defined exactly as expected as

$$\mathcal{AB} = \{(x, z) \in \mathcal{H}_1 \oplus \mathcal{H}_3 : (x, y) \in \mathcal{B}, (y, z) \in \mathcal{A} \text{ for some } y \in \mathcal{H}_2\}.$$

https://doi.org/10.1515/9783110563238-003

Condition (3) implies its more familiar operator version $U_1SU_1^* = M_t$, and here U_1 denotes the unitary part of $U = U_1 \oplus 0$, and, as always, S denotes the operator part of $\mathcal{S}$. In fact, the two versions are equivalent because $U_1^*f = U^*f$ for all f and the multi-valued part of $\mathcal{S}$ (that is missing from S) makes no extra contribution to $U\mathcal{S}U^*$ because its members are annihilated by U.

Before we proceed, I also take the opportunity to remind the reader that the *discrete spectrum* σ_d of a self-adjoint operator is defined as the set of all isolated (in the spectrum) eigenvalues of finite multiplicity. For a canonical system, the multiplicity is at most 2, so σ_d can also be described as the set of all isolated points of σ. It need not be closed and thus is not really a spectrum. The *essential spectrum* can be defined as $\sigma_{\text{ess}} = \sigma \setminus \sigma_d$.

Weyl's theorem says that the essential spectrum is invariant under compact perturbations. The same statement is obtained if the resolvents $(S_j - i)^{-1}$ of two operators differ by a compact perturbation, and it is in this form that the result is usually applied to spectral problems for differential equations.

3.2 The resolvent as an integral operator

We define the *resolvent* of a self-adjoint relation $\mathcal{S}$ in the same way as for operators as $(\mathcal{S} - z)^{-1}$, with $z \in \mathbb{C}$. It has the same basic properties as in the operator case, except that the multi-valued part of $\mathcal{S}$ will give it a kernel.

Theorem 3.1. *Let $\mathcal{S}$ be a self-adjoint relation, and $z \notin \sigma(\mathcal{S})$. Then $(\mathcal{S} - z)^{-1}$ is a bounded normal operator with $N((\mathcal{S} - z)^{-1}) = \mathcal{S}(0)$.*

Proof. If $z \notin \sigma(\mathcal{S})$, then certainly $N(\mathcal{S} - z) = 0$, so it is clear that $(\mathcal{S} - z)^{-1}$ is an operator. We then use the decomposition $\mathcal{H} = \overline{D(\mathcal{S})} \oplus \mathcal{S}(0)$ into reducing subspaces and the induced decomposition $\mathcal{S} = \{(x, Sx)\} \oplus (0, \mathcal{S}(0))$ from Theorem 2.11. This shows that (in operator notation) $(\mathcal{S} - z)^{-1} = (S - z)^{-1} \oplus 0$, which is bounded and normal as an orthogonal sum of two such operators. The claim on $N((\mathcal{S} - z)^{-1})$ also is an immediate consequence of this representation. □

We now show that the resolvent of a self-adjoint realization of a canonical system is an integral operator, and we give a formula for its kernel. We need to set up some notation. Fix a self-adjoint realization $\mathcal{S}$; if both endpoints are limit circle, then we also assume that the boundary conditions are separated. We say that a solution u of $Ju' = -zHu$ is in $D(\mathcal{S})$ near an endpoint t, $t = a, b$, if $u \in L_H^2$ near t and u satisfies the boundary condition at t if there is one. Or, more explicitly, if we have limit point case at t, then $u \in D(\mathcal{S})$ near t just means that $u \in L_H^2$ there; if t is in the limit circle case, then this condition is automatic for $z \notin \mathbb{R}$ by Theorem 2.31 (and in fact it follows for all $z \in \mathbb{C}$, as we will see in the next section), so now $u \in D(\mathcal{S})$ near t means that u satisfies the boundary condition at t. We need to clarify what exactly we mean by this, and we proceed as follows: the boundary condition will be of the form $(f_0^* J h_0)(t) = 0$, for some

$(f,g) \in \mathcal{S}$. Since $u \in L^2_H$ near t now, Lemma 2.7 shows that there are $(h,k) \in \mathcal{T}$ with $h_0 = u$ near t. Thus existence of the limit $(f_0^* Ju)(t)$ is guaranteed, by Theorem 2.8, and what we ask for is that its value be zero.

Lemma 3.2. *Let $\mathcal{S}$ be as above and let $z \in \mathbb{C} \setminus \mathbb{R}$. Then at each endpoint, there is a unique, up to a constant multiple, non-trivial solution u of $Ju' = -zHu$ that lies in $D(\mathcal{S})$ near that endpoint.*

Proof. This is obvious at a limit point endpoint. At a limit circle endpoint, there is certainly some non-trivial solution u that satisfies the boundary condition since the candidate u's come from a two-dimensional space and we only need to make one functional vanish.

If all solutions satisfied the boundary condition, then at least one of these would also be in $D(\mathcal{S})$ near the other endpoint, by the same argument, but that would lead to a non-real eigenvalue z of the self-adjoint operator S, which is impossible. □

Theorem 3.3. *Let $\mathcal{S}$ be as above, $z \in \mathbb{C} \setminus \mathbb{R}$, and for each endpoint $t = a, b$, pick a non-trivial solution u_t of $Ju' = -zHu$ that lies in $D(\mathcal{S})$ near t. Normalize these so that $W(u_a, u_b) = -1$. Let*

$$G(x,y;z) = \begin{cases} u_a(x)u_b^t(y), & x \le y, \\ u_b(x)u_a^t(y), & x > y. \end{cases}$$

Then, for all $f \in L^2_H(a,b)$,

$$((\mathcal{S} - z)^{-1} f)(x) = \int_a^b G(x,y;z)H(y)f(y)\,dy. \tag{3.1}$$

Recall that $W(u,v) = u^t Jv$ denotes the (constant) Wronskian of u, v. We know that u_a, u_b cannot be linearly dependent, or otherwise we would again obtain a non-real eigenvalue, so $W(u_a, u_b) \neq 0$, and then multiplying u_a or u_b by a suitable constant will make $W = -1$.

The notation on the left-hand side of (3.1) is intuitive but somewhat sloppy from a formal point of view: I really mean the *operator*, applied to f, so I have identified the relation with the operator. The integral on the right-hand side converges for all $x \in (a,b)$ if $f \in L^2_H(a,b)$, and the function $g(x)$ it defines represents $(\mathcal{S} - z)^{-1} f$. More is true: $g(x)$ is the unique absolutely continuous representative that satisfies the boundary conditions (if any) and solves $Jg' = -zHg - Hf$, as we will see in the proof.

The integral kernel G is also called the *Green function*. It does depend on z because the solutions u_a, u_b do.

Proof. One could try to derive the formula for G systematically from the variation of constants formula, but it is much easier to just check that it works. Since

$$\int_a^b GHf\,dy = u_b(x) \int_a^x u_a^t(y)H(y)f(y)\,dy + u_a(x) \int_x^b u_b^t(y)H(y)f(y)\,dy$$

and each u_t is in L^2_H near its endpoint, it is clear that the integral converges for all $x \in (a,b)$ and defines an absolutely continuous function of x. Differentiation of this formula then shows that $g(x) \equiv \int GHf\,dy$ solves $Jg' = -zHg + J(u_b u_a^t - u_a u_b^t)Hf$, and now a straightforward calculation confirms that

$$J(u_b(x)u_a^t(x) - u_a(x)u_b^t(x)) = -1 \ (\in \mathbb{C}^{2\times 2}),$$

so $Jg' = -zHg - Hf$. Now if f has compact support in (a,b), then $g \in L^2_H(a,b)$, so $g \in D(\mathcal{T})$, and g satisfies the boundary conditions. This follows because these properties could only fail near the endpoints, but for f of compact support, we have $g(x) = c_a u_a(x)$ near a and $g(x) = c_b u_b(x)$ near b. Thus $g \in D(\mathcal{S})$ in this case, and hence $(g,f) \in \mathcal{S} - z$. Since $(\mathcal{S} - z)^{-1}$ is an operator, this suffices to identify g as $(\mathcal{S} - z)^{-1}f$.

The compactly supported fs form a dense subspace of $L^2_H(a,b)$, and $(\mathcal{S} - z)^{-1}$ is a bounded operator, so $(\mathcal{S} - z)^{-1}f$ for a general $f \in L^2_H(a,b)$ may now be obtained by approximation. However, if $f_n \to f$, then the right-hand sides of (3.1) also converge pointwise because $G(x,y;z) \in L^2_H(a,b)$ as a function of y for each fixed x. Moreover, the norm limit $(\mathcal{S} - z)^{-1}f_n$ will become a pointwise limit after passing to a suitable subsequence. More precisely, this will allow us to assume that

$$H(x)((\mathcal{S} - z)^{-1}f_n)(x) \to H(x)((\mathcal{S} - z)^{-1}f)(x)$$

for almost every $x \in (a,b)$. Thus $H(x)g(x) = H(x)((\mathcal{S} - z)^{-1}f)(x)$ almost everywhere, that is, $g(x)$ does represent $(\mathcal{S} - z)^{-1}f$.

To prove the additional claims, made explicit in the comments following the theorem, that $g(x)$ is the unique absolutely continuous representative that satisfies the boundary conditions (if any) and solves $Jg' = -zHg - Hf$, recall that we already showed that our g is a solution of this equation. So if it was not the right one itself, then $g(x) + u(x)$ would be, for some solution u of $Ju' = -zHu$. However, since $g + u$ must then still represent the same Hilbert space element as g, this u would have to satisfy $Hu = 0$ almost everywhere, so $u = 0$ after all. □

Theorem 3.4. *Assume limit circle case at both endpoints, and consider a self-adjoint realization $\mathcal{S}$ with separated boundary conditions. Then $(\mathcal{S} - i)^{-1}$ is a Hilbert–Schmidt operator. As a consequence, the spectrum $\sigma(\mathcal{S}) = \{E_n\}$ is purely discrete, and $\sum \frac{1}{1+E_n^2} < \infty$.*

This is of course also true for general, possibly coupled, boundary conditions and in fact follows from Theorem 3.4 as stated because a change of boundary condition amounts to a finite rank perturbation of the resolvent.

Proof. This follows from Theorem 3.3. With the limit circle assumptions, $u_a, u_b \in L^2_H(a,b)$ (not just near their own endpoints), and this will make the kernel of $(\mathcal{S} - i)^{-1}$ square integrable. To check this formally, the isometric embedding $V : L^2_H(a,b) \to L^2_I(a,b)$, $(Vf)(x) = H^{1/2}(x)f(x)$ is useful. It transforms $(\mathcal{S} - i)^{-1}$ into an integral operator on a subspace of L^2_I with the new kernel $K(x,y) = H^{1/2}(x)u_a(x)u_b^t(y)H^{1/2}(y)$ for

$x \leq y$ and a similar formula for $x \geq y$. The elements of $K(x,y)$ are now square integrable functions with respect to Lebesgue measure, so the transformed operator is Hilbert–Schmidt, and thus so is the original one, being unitarily equivalent to it. □

3.3 A limit point/limit circle criterion

For the classical equations (Schrödinger, Jacobi, etc.), there are rather precise results on the limit point/limit circle alternative, but no characterization in terms of the behavior of the coefficient functions near the endpoint under consideration. For canonical systems, we have the following beautiful result.

Theorem 3.5. *An endpoint is in the limit point case if and only if $H \notin L^1$ near that endpoint.*

Since $H(x) \geq 0$, this condition is equivalent to $\operatorname{tr} H \notin L^1$.

One direction of the Theorem is obvious: if $H \in L^1$ near an endpoint, then that endpoint is regular and thus in the limit circle case. This remark also gives us a rather striking reformulation of Theorem 3.5.

Corollary 3.6. *A limit circle endpoint is regular.*

These results do actually become quite plausible if we recall that $H(x)$ should really be viewed as information about solutions at $z = 0$, as suggested at the end of Section 1.3. Then Theorem 3.8(b), (c) below provides a rather convincing explanation of Theorem 3.5 and its corollary. In the formal argument, we will obtain these results as a consequence of the following general fact.

Lemma 3.7. *Let $\mathcal{T}_0$ be a closed symmetric relation with equal deficiency indices $\gamma_+ = \gamma_- = \gamma$ and write $\mathcal{T} = \mathcal{T}_0^*$. Suppose that*

$$\gamma(t) \equiv \dim N(\mathcal{T} - t) \neq \gamma.$$

Then $t \in \sigma(\mathcal{S})$ for all self-adjoint realizations $\mathcal{S}$.

Proof. Consider the regularity domains of $\mathcal{T}_0$ and its extensions. Clearly, if $\mathcal{S} \supseteq \mathcal{T}_0$, then $\Gamma(\mathcal{S}) \subseteq \Gamma(\mathcal{T}_0)$, and if $\mathcal{S}$ is self-adjoint, then $\mathbb{C} \setminus \sigma(\mathcal{S}) \subseteq \Gamma(\mathcal{S})$: off the spectrum, the resolvent $(\mathcal{S} - z)^{-1}$ will be a bounded operator, and this is the condition that defined the regularity domain. Finally, $\gamma(z) = \gamma$ on $z \in \Gamma(\mathcal{T}_0)$ by Theorem 2.31. When these observations are combined, the lemma follows. □

Proof of Theorem 3.5. By our preliminary discussion, it suffices to prove the following: if b is in the limit circle case, then $\operatorname{tr} H \in L^1(c,b)$ for some (or every) interior point $c \in (a,b)$. We apply Lemma 3.7 to this problem on (c,b), which has deficiency index $\gamma = 2$. It is not possible here to have at most one $L^2_H(c,b)$ solution of $Ju' = -tHu$ and, at the same time, $t \in \sigma(\mathcal{S})$ for all self-adjoint realizations $\mathcal{S}$. This follows because the spectrum is purely discrete, by Theorem 3.4, so t would in fact have to be an eigenvalue of all $\mathcal{S}$, but we only have at most one candidate eigenvector to work with, namely the solution u from above, and this satisfies $u(c) \neq 0$ (or else $u \equiv 0$), so satisfies only one boundary condition at c. So we conclude that $\gamma(t) = 2$ for all t.

In particular, we can take $t = 0$ here, and then the solutions u of $Ju' = -0 \cdot Hu$ are simply constants, so $u_1(x) = e_1$ and $u_2(x) = e_2$ are both in $L^2_H(c, b)$. Thus

$$\int_c^b \operatorname{tr} H(x)\, dx = \|u_1\|^2_{L^2_H(c,b)} + \|u_2\|^2_{L^2_H(c,b)} < \infty. \qquad \square$$

Lemma 3.7 is of considerable independent interest, and for a canonical system, we can elaborate some more on this theme.

Theorem 3.8. *Consider a canonical system with deficiency index γ and recall that $\gamma(z)$ was defined as the number of linearly independent $L^2_H(a, b)$ solutions of $Ju' = -zHu$.*
(a) *If $\gamma = 2$, then $\gamma(z) = 2$ for all $z \in \mathbb{C}$.*
(b) *If $\gamma = 1$, then $\gamma(z) \leq 1$ for all $z \in \mathbb{C}$, and if $\gamma(t) = 0$, then $t \in \sigma_{ess}(S)$ for all self-adjoint realizations $\mathcal{S}$.*
(c) *If $\gamma = 0$, then $\gamma(t) \leq 1$ for all $t \in \mathbb{R}$.*

Proof. Part (a) was essentially established in the previous proof. Now that Theorem 3.5 has been proved, we also obtain the statement as an immediate consequence of Corollary 3.6: if $\gamma = 2$, then both endpoints are regular, and this makes it obvious that $\gamma(z) = 2$ for all $z \in \mathbb{C}$.

(b) We only need to show that $\gamma(t) \leq 1$ for all $t \in \mathbb{R}$. The second claim is immediate from Lemma 3.7 since t is certainly not an eigenvalue of any $\mathcal{S}$ if $\gamma(t) = 0$. So suppose then we had $\gamma(t) = 2$ for some $t \in \mathbb{R}$. By Corollary 3.6, one endpoint (let's say a) is regular while the other is limit point. We can now consider $\mathcal{T}_1 = \mathcal{T}_0 + \mathcal{N}_t$, with, as above, $\mathcal{N}_t = \{(u, tu) \in \mathcal{T}\}$; this is a two-dimensional extension of $\mathcal{T}_0$ because a $(u, tu) \in \mathcal{N}_t$ will be in $\mathcal{T}_0$ only if $u_0(a) = 0$, but since u_0 must also solve $Ju' = -tHu$, this makes $u = 0$. Moreover, $\mathcal{T}_1$ is easily seen to be symmetric. This is impossible because $\mathcal{T}_0$ had deficiency index 1, so there cannot be a two-dimensional symmetric extension.

In part (c), we can again make an interior $c \in (a, b)$ a new endpoint and then apply the argument from part (b) to (c, b). $\square$

3.4 Weyl theory

In this section, we introduce what will turn out to be a key tool in the spectral analysis of canonical systems, the Titchmarsh–Weyl m function. This will provide a link between the solutions of

$$Ju' = -zHu$$

and spectral data; in fact, we can use the m function to construct these spectral data in the first place.

The whole method requires the existence of at least one regular endpoint. For convenience, we take specifically $(a, b) = (0, \infty)$, with 0 regular. I will frequently write the basic interval as $[0, \infty)$, to emphasize this fact, that 0 is a regular endpoint, but I am under no obligation of doing so and could also continue to write $(0, \infty)$.

We will be mostly interested in the case when we have limit point case at infinity. By Theorem 2.25(b), the self-adjoint realizations are then obtained by imposing a boundary condition at $x = 0$, which, in its general form, is given by

$$u_1(0) \sin \alpha - u_2(0) \cos \alpha = 0.$$

We will develop the theory only for the specific choice $\alpha = 0$, so the boundary condition becomes $u_2(0) = 0$. What we would have obtained from general boundary conditions is contained in this special case, thanks to the transformations $H_A = A^{-1t} H A^{-1}$, as I will discuss in more detail later, in Section 3.6.

We now introduce an intermediate point $L > 0$ and consider the problem on $[0, L]$ and its self-adjoint realizations with separated boundary conditions, as described in part (c) of Theorem 2.25. At $x = 0$, we impose the same boundary condition $u_2(0) = 0$ as for the half line problem. The boundary condition at $x = L$, on the other hand, we keep general, so consider the problem on $[0, L]$ with boundary conditions

$$u_2(0) = 0, \quad u_1(L) \sin \beta - u_2(L) \cos \beta = 0.$$

The basic assumption that our canonical systems never consist of a single singular interval applies to the choice of L also; in other words, if $(0, \infty)$ starts with a singular interval $(0, d)$, then it will be understood that $L > d$.

The m function $m : \mathbb{C}^+ \to \mathbb{C}$, with $\mathbb{C}^+ = \{z \in \mathbb{C} : \operatorname{Im} z > 0\}$, for this problem is now defined as follows: take a non-trivial solution $f = f(x, z)$ of $Jf' = -zHf$ that satisfies the boundary condition at L and then set

$$m = m_L^{(\beta)}(z) = \frac{f_1(0, z)}{f_2(0, z)}. \tag{3.2}$$

In fact, it is better to write this as $m = f(0, z)$, with the understanding that a non-zero vector v is identified with the point $v_1/v_2 \in \mathbb{C}_\infty$ on the Riemann sphere $\mathbb{C}_\infty = \mathbb{C} \cup \{\infty\}$. This also takes care of the concern that $f_2(0, z)$ might be zero (though that does actually not happen for $z \in \mathbb{C}^+$). Or, equivalently, we can say that we view $f(0, z)$ as a point in projective space $\mathbb{CP}^1$ (the one-dimensional subspaces of $\mathbb{C}^2$), and then we identify $\mathbb{CP}^1 \cong \mathbb{C}_\infty$ in the way indicated.

Whichever interpretation we use, note that m is well defined since f is determined by the boundary condition at L up to a constant multiple, and this undetermined constant drops out when we take the quotient f_1/f_2. More precisely, $f(L, z)$ must be a multiple of $e_\beta = (\cos\beta, \sin\beta)^t$, and since solutions are updated by the transfer matrix, this gives

$$m_L^{(\beta)}(z) = T^{-1}(L;z)e_\beta, \quad T(L;z) \equiv T(L,0;z). \tag{3.3}$$

The natural action of an invertible matrix on vectors and then on projective space induces a corresponding action on the Riemann sphere, by using this identification of $\mathbb{CP}^1$ with $\mathbb{C}_\infty$. If we write it out, we see that the action is by linear fractional transformations

$$\begin{pmatrix} a & b \\ c & d \end{pmatrix} w = \frac{aw+b}{cw+d}, \qquad w \in \mathbb{C}_\infty.$$

From the above remarks, it is clear that this indeed defines a group action of $\mathrm{SL}(2,\mathbb{C})$ on $\mathbb{C}_\infty$ (and also of $\mathrm{GL}(2,\mathbb{C})$, but our matrices are transfer matrices and thus have determinant 1). With these agreements in place, (3.3) could also have been written as

$$m_L^{(\beta)}(z) = T^{-1}(L;z)\cot\beta, \tag{3.4}$$

and here we use the additional, almost self-explanatory convention that a matrix followed by a number refers to this action by linear fractional transformations.

This flexible and convenient notation, which allows us to move back and forth between vectors and points on the Riemann sphere almost at will, will be used extensively in the sequel. Occasionally, you will have to exercise some care in interpreting formulae. For example, if $v \in \mathbb{C}^2$, $v \neq 0$, then $-v$ could mean two things in addition to its literal interpretation as the vector $-v$: it could refer to the negative of the number represented by v, or it could mean the number represented by $-v$ (which is the number represented by v, making this second version unlikely). The convenience, elegance, and transparency gained by using these notations will more than compensate for this slight potential for confusion.

One important consequence of (3.4), combined with Theorem 1.2, will be that m maps the upper half plane back to itself, and it is useful to state the underlying general fact separately. But first let me introduce some terminology.

Definition 3.2. A *Herglotz function* is a holomorphic map $F : \mathbb{C}^+ \to \mathbb{C}^+$. A *generalized Herglotz function* is a holomorphic map $F : \mathbb{C}^+ \to \mathbb{C}_\infty$ taking values in $\overline{\mathbb{C}^+}$.

In other words, the generalized Herglotz functions are $F(z) \equiv a$, $a \in \mathbb{R}_\infty$, and the genuine Herglotz functions.

Lemma 3.9. *Let $A \in \mathrm{SL}(2,\mathbb{C})$. Then $z \mapsto Az$ is a Herglotz function if and only if $i(J - A^*JA) \geq 0$.*

The reader can also check that $J - A^*JA = 0$ precisely if $A \in \mathrm{SL}(2,\mathbb{R})$, and now recall that these are exactly the automorphisms of $\mathbb{C}^+$.

Proof. One direction is obvious since $-iv^*Jv$ is a positive multiple of $\operatorname{Im} v$ for any nonzero vector v. So assume now that $z \mapsto Az$ is a Herglotz function. Then A maps $\mathbb{R}_\infty$ to a circle on the Riemann sphere which is contained in $\overline{\mathbb{C}^+}$, so in the plane, this becomes either a genuine circle or a line parallel to the real axis in $\overline{\mathbb{C}^+}$. In either case, we can compose with a shift so that now the bottom of this circle touches 0, so $t \mapsto 0$ for some $t \in \mathbb{R}_\infty$. Then we can also shift t so that now $t' = 0$ gets mapped to 0. In other words, we consider

$$B = \begin{pmatrix} 1 & -w \\ 0 & 1 \end{pmatrix} A \begin{pmatrix} 1 & t \\ 0 & 1 \end{pmatrix},$$

and here $w \in \mathbb{C}^+ \cup \mathbb{R}$, $t \in \mathbb{R}$ (this excludes the case $t = \infty$, which I leave to the reader); B is still a Herglotz function. So now the original matrix can be written as

$$A = \begin{pmatrix} 1 & w \\ 0 & 1 \end{pmatrix} B \begin{pmatrix} 1 & -t \\ 0 & 1 \end{pmatrix} \equiv CBD.$$

To establish that $-iA^*JA \geq -iJ$, it suffices to show that $-iX^*JX \geq -iJ$ for the individual factors $X = C, B, D$. This is obvious for D, with equality, since $D \in SL(2, \mathbb{R})$, and C, by a calculation.

Since $B0 = 0$, this matrix is of the form

$$B = \begin{pmatrix} a & 0 \\ b & a^{-1} \end{pmatrix}.$$

Let $z = x + iy \in \mathbb{C}$, set $v = (z, 1)^t$, and look at

$$\frac{-i}{2} v^* B^* JBv = \alpha(x^2 + y^2) + \beta y + \gamma x \equiv F(x, y), \tag{3.5}$$

and here

$$\alpha = \operatorname{Im} a\overline{b}, \quad \beta = \operatorname{Re} a/\overline{a}, \quad \gamma = \operatorname{Im} a/\overline{a}.$$

As we observed earlier, the left-hand side of (3.5) computes a positive multiple of $\operatorname{Im} Bz$, so $F(x, y) \geq 0$ for $x \in \mathbb{R}$, $y \geq 0$. This implies that $\alpha \geq 0$ and also that $\gamma = 0$, which in turn makes $\beta = \pm 1$, but only $\beta = 1$ is consistent with $F \geq 0$. Thus

$$F(x, y) = \alpha(x^2 + y^2) + y \geq y = \frac{-i}{2} v^* Jv,$$

as we wanted to show. □

Theorem 3.10. *The m function $m_L^{(\beta)}(z)$ is a Herglotz function and has a meromorphic extension, with all poles on the real line.*

Proof. We first show that $m(z) \equiv a \in \mathbb{R}_\infty$ is not possible. In this case, the solution f of $Jf' = -zHf$ from (3.2) would be a multiple of a real vector $v \in \mathbb{R}^2$ at $x = 0$, so would satisfy some boundary condition α there. This would then seem to make $z \notin \mathbb{R}$ an eigenvalue of the self-adjoint realization $\mathcal{S} = \mathcal{S}_{\alpha,\beta}$ that we obtain from these boundary conditions. Note that since $f \in D(\mathcal{S})$, the element $(f, zf) \in \mathcal{S}$ is the correct one if we want to extract the operator part S, so such an f would indeed satisfy $Sf = zf$. We have two conceivable ways out of this: the function f could represent the zero element of $L^2_H(0,L)$, or $(0,L)$ could be a singular interval, so that our theory does not apply directly. However, we explicitly ruled out the second scenario, and in the first case, it also follows that then $(0,L)$ must be a single singular interval, of type $\beta^\perp$.

In the original definition (3.2), f can be the solution with $f(L,z) = e_\beta$. So m is indeed meromorphic; here we use the information gained in the previous step, that $m \equiv \infty$ is not possible. Since f is real on the real line, the non-real poles of m come in complex conjugate pairs. By combining (3.4), Lemma 3.9, and Theorem 1.2, we see that m is a generalized Herglotz function and then also a genuine Herglotz function by the first step again. This property then rules out poles on $\mathbb{C}^+$, so all poles are real. □

We now introduce the key objects of the whole analysis.

Definition 3.3. For $z \in \mathbb{C}^+$, we define the *Weyl circle* $\mathcal{C}(L;z)$ and the *Weyl disk* $\mathcal{D}(L;z)$ as the sets

$$\mathcal{C}(L;z) = \{T^{-1}(L;z)q : q \in \mathbb{R}_\infty\},$$
$$\mathcal{D}(L;z) = \{T^{-1}(L;z)q : q \in \overline{\mathbb{C}^+}\}.$$

The closure of $\mathbb{C}^+$ in the definition of $\mathcal{D}$ is taken in the Riemann sphere, so includes ∞.

These definitions are in an obvious way motivated by our previous discussion. Indeed, the Weyl circle

$$\mathcal{C}(L;z) = \{m_L^{(\beta)}(z) : \beta \in [0,\pi)\}$$

just collects the possible values of m when we vary the boundary condition at L.

Recall that a linear fractional transformation sends circles to circles on $\mathbb{C}_\infty$, and a circle not containing ∞ is a circle in the plane $\mathbb{C}$, while a circle through ∞ becomes a line in the plane. So the Weyl circle and disk are appropriately named: they are a circle and a disk, respectively, at least on the sphere. If we move back to the plane, then the Weyl circle might be a line. In Theorem 3.12, we will see that $\mathcal{C}$ is a genuine circle. Moreover, $\mathcal{D}$ is its interior (not exterior) and thus a disk in the plane also.

Lemma 3.11. *Suppose that*

$$A = \begin{pmatrix} a & b \\ c & d \end{pmatrix} \in \mathrm{SL}(2,\mathbb{C})$$

and define $C := \{Aq : q \in \mathbb{R}_\infty\}$. *Then the radius* R *of* C *is given by*

$$\frac{1}{R} = 2|\operatorname{Im} \bar{c}d|.$$

By our discussion preceding the Lemma, C is a circle on the Riemann sphere $\mathbb{C}_\infty$, which corresponds to a circle or a line in the plane $\mathbb{C}$. The radius R that the Lemma computes is taken in the plane, and if C is a line there, then we define its radius to be $R = \infty$. With this interpretation, the formula for R works in all cases. In particular, it shows that C is a line if and only if $\operatorname{Im} \bar{c}d = 0$.

Proof. We decompose the linear fractional transformation A into translations, dilations, and an inversion. Consider first the case $c \neq 0$. Then we can write

$$A = \begin{pmatrix} 1 & a/c \\ 0 & 1 \end{pmatrix} \begin{pmatrix} 1/c & 0 \\ 0 & c \end{pmatrix} J \begin{pmatrix} 1 & d/c \\ 0 & 1 \end{pmatrix},$$

and now we just keep track of where $\mathbb{R}_\infty$ gets sent (in the plane) under this sequence of maps. The rightmost matrix acts as a translation by d/c, and this maps $\mathbb{R}$ onto the line parallel to $\mathbb{R}$ at height $\operatorname{Im} d/c$, and let us also assume for now that this is non-zero. Next, the inversion $Jz = -1/z$ sends this to a circle of radius $r = 1/(2|\operatorname{Im} d/c|)$ and with $\pm ir/2$ as its center. The next matrix acts by multiplication by $1/c^2$, so this multiplies the radius by $1/|c|^2$, and then the final matrix acts again as a translation and will not affect the radius. We obtain the stated formula.

If $c \neq 0$, but $\operatorname{Im} d/c = 0$, then this argument still works (in fact, it simplifies), and we see that $\mathbb{R}$ gets sent to a line. The formula for $1/R$ is correct in this case also.

The case $c = 0$ is equally straightforward. In fact, it suffices to observe that then $A\infty = \infty$, so $A\mathbb{R}$ must be a line. □

We introduce one more piece of notation: we will denote by u, v the functions $u(x,z) = T(x;z)e_1$, $v(x,z) = T(x;z)e_2$. So these are the solutions of $Ju' = -zHu$ with the initial values e_1 and e_2, respectively. The solution u satisfies the boundary condition at $x = 0$, and

$$f(x,z) = v(x,z) + m_L^{(\beta)}(z)u(x,z) = T(x;z)\begin{pmatrix} m_L^{(\beta)}(z) \\ 1 \end{pmatrix}$$

is a solution that satisfies the boundary condition β at $x = L$. In fact, it is the unique solution of the form $v + Mu$ with this property, and this gives still another characterization of m.

Theorem 3.12. *The Weyl disks satisfy* $\mathcal{D}(L;z) \subseteq \mathbb{C}^+$ *for* $z \in \mathbb{C}^+$, *and they are nested:*

$$\mathcal{D}(L_2;z) \subseteq \mathcal{D}(L_1;z) \quad (L_2 \geq L_1).$$

The radius $R = R(L;z)$ of $\mathcal{D}(L;z)$ is given by

$$\frac{1}{R} = 2\operatorname{Im} z \int_0^L u^*(x,z)H(x)u(x,z)\,dx.$$

Proof. The first statement follows exactly as in the proof of Theorem 3.10 by combining Lemma 3.9 and Theorem 1.2. More precisely, this will show that $\mathcal{D}(L;z) \subseteq \overline{\mathbb{C}^+}$, but then we also know that $\mathcal{C}(L;z)$ cannot contain points from $\mathbb{R}_\infty$.

That the disks are nested is obvious from their definition because if $L_2 \geq L_1$, then

$$\mathcal{D}(L_2;z) = \{T^{-1}(L_2)q : q \in \overline{\mathbb{C}^+}\} = \{T^{-1}(L_1)T^{-1}(L_2,L_1)q : q \in \overline{\mathbb{C}^+}\},$$

and this is a subset of $\mathcal{D}(L_1;z)$ since $T^{-1}(L_2,L_1)$ also maps $\overline{\mathbb{C}^+}$ back to itself.

Since

$$T^{-1}(L;z) = \begin{pmatrix} v_2(L,z) & -v_1(L,z) \\ -u_2(L,z) & u_1(L,z) \end{pmatrix},$$

an application of Lemma 3.11 shows that

$$\pm\frac{1}{R} = i(\overline{u_2(L)}u_1(L) - u_2(L)\overline{u_1(L)}) = iu^*(L)Ju(L).$$

Since $(u^*(x)Ju(x))' = (\overline{z} - z)u^*Hu$ and $u^*Ju = 0$ at $x = 0$, this can be rewritten as in the theorem. □

A different description of $\mathcal{C}$ and $\mathcal{D}$ is often useful. For an arbitrary number $M \in \mathbb{C}$, let

$$f_M(x,z) = T(x;z)\begin{pmatrix} M \\ 1 \end{pmatrix} = v(x,z) + Mu(x,z).$$

Theorem 3.13. *Let $M \in \mathbb{C}$, $z = x + iy \in \mathbb{C}^+$. Then $M \in \mathcal{D}(L;z)$ if and only if*

$$y\int_0^L f_M^*(x)H(x)f_M(x)\,dx \leq \operatorname{Im} M, \tag{3.6}$$

and $M \in \mathcal{C}(L;z)$ if and only if this holds with equality.

Proof. This follows from the by now familiar identity

$$2iy\int_0^L T^*(x)H(x)T(x)\,dx = J - T^*(L)JT(L),$$

which we (as usual) obtain by working out $(T^*JT)'$. Multiply by w^* and w, with $w = (M,1)^t$, from the left and right, respectively. Then the left-hand side becomes the left-hand side of (3.6), with an extra factor of $2i$. On the right-hand side, $w^*Jw = 2i\operatorname{Im} M$

and, similarly, $w^*T^*JTw = 2i\mathrm{Im}\, Tw$. So (3.6) is now seen to be equivalent to the requirement that $TM \in \overline{\mathbb{C}^+}$ or $M \in T^{-1}\overline{\mathbb{C}^+}$, and this is how we defined $\mathcal{D}$.

Essentially the same argument, with a small modification in the final few steps, gives the claim about $\mathcal{C}$. □

Since this calculation did not depend on any of our previous results in this section, (3.6) also provides an alternative and in fact much quicker proof of the Herglotz property of m. What we did seems conceptually more transparent. More importantly, in this way we saw that the whole theory only depends on the properties of the transfer matrix from Theorem 1.2, not on the form of the differential equation itself. This will become important in Chapter 4, when we study abstract transfer matrices.

Since the Weyl disks $\mathcal{D}(L;z)$ are nested and compact, they will converge to a nonempty limiting object

$$\mathcal{D}(z) \equiv \bigcap_{L>0} \mathcal{D}(L;z)$$

as $L \to \infty$, and $\mathcal{D}(z)$ is either a point or a disk.

Now we make the important observation that if $M \in \mathcal{D}(z)$, then also

$$y \int_0^\infty f_M^*(x)H(x)f_M(x)\, dx \le \mathrm{Im}\, M; \tag{3.7}$$

this is an immediate consequence of (3.6), which is valid for all $L > 0$ if $M \in \mathcal{D}(z)$.

Theorem 3.14. (a) *If we have limit point case at infinity, then $\mathcal{D}(z)$ is a point for all $z \in \mathbb{C}^+$. The unique $m = m(z) \in \mathcal{D}(z)$ satisfies*

$$y \int_0^\infty f_m^*(x)H(x)f_m(x) = \mathrm{Im}\, m(z). \tag{3.8}$$

(b) *If we have limit circle case at infinity, then $\mathcal{D}(z)$ is a disk for all $z \in \mathbb{C}^+$.*

This finally justifies, in grand style, the terminology that was introduced in Definition 2.6.

We can say much more about the $m \in \mathcal{D}$ in case (b) also, but it is really not necessary to make any of this explicit. The limit circle endpoint ∞ is really regular, so we are back in the case we started the whole discussion with, and everything we said about the problems on $[0, L]$ now applies to $[0, \infty)$ as well. The limit circle $\mathcal{C}(z) = \partial\mathcal{D}(z)$ is

$$\mathcal{C}(z) = \{T^{-1}(\infty; z)q : q \in \mathbb{R}_\infty\},$$

which is the collection of m functions we obtain when we vary the boundary condition at ∞, and we then obtain $\mathcal{D}(z)$ by filling up the interior of this circle. All the previous results also apply to the limit circle and disk for the trivial reason that these are of exactly the same type as the Weyl circles and disks of the problems on $[0, L]$.

Proof. (a) By (3.7), any $M \in \mathcal{D}(z)$ contributes an $L^2_H(0,\infty)$ solution of $Ju' = -zHu$, and $f_M, f_{M'}$ are linearly independent if $M \neq M'$. However, we know that $\gamma(z) = \gamma = 1$ for all $z \in \mathbb{C}^+$. Thus $\mathcal{D}(z)$ cannot contain more than one point.

To prove (3.8), recall that since $f_m \in L^2_H$, we have $(f_m, zf_m) \in \mathcal{T}$, so $f_m^*(L)Jf_m(L) \to 0$ as $L \to \infty$ by Theorem 2.24. Note here that f_m itself *is* the representative determined by (f_m, zf_m) (what we used to denote by $f_{m,0}$).

Now return to the proof of Theorem 3.13: We saw there that the discrepancy between the two sides of (3.6) was a multiple of $f_M^*(L)Jf_M(L)$, so taking the limit $L \to \infty$ in our situation will produce (3.8).

(b) This statement is trivial and was already discussed at greater length following the formulation of the theorem; it is included for contrast and completeness. If we have limit circle case at infinity, then this endpoint is really regular and $T(L;z)$ approaches the limit $T(\infty;z)$ as $L \to \infty$, and when $L \to \infty$ approaches a regular endpoint, then the Weyl disks $\mathcal{D}(L;z)$ simply shrink to the one of $L = \infty$, so the limiting object is no different from the approximating ones. □

In the proof of part (a), we used the constancy of the deficiency index in the argument. This could be turned around. Rather standard complex analysis methods can be used to show that as soon as $\mathcal{D}(z)$ is a point for a single $z = z_0 \in \mathbb{C}^+$, it will be a point for all $z \in \mathbb{C}^+$. This then provides an alternative proof of the constancy of the deficiency index for canonical systems.

In the limit point case, the restriction of $\mathcal{T}$ by the boundary condition $u_2(0) = 0$ defines a self-adjoint realization, and we define its m function $m : \mathbb{C}^+ \to \mathbb{C}^+$ as the unique point $m(z) \in \mathcal{D}(z)$.

Theorem 3.15. *Assume limit point case at infinity. Then $m(z)$ is a Herglotz function. If f denotes the (unique, up to a factor) non-trivial $L^2_H(0,\infty)$ solution of $Ju' = -zHu$, then $m(z) = f(0,z)$.*

In other words, (3.2) also works for the half line m function $m(z)$, if we replace the original requirement that f satisfy the boundary condition at the right endpoint by the condition that $f \in L^2_H$ there. Somewhat more elegantly from a formal point of view, we can define f as the solution that is in $D(\mathcal{S})$ near the right endpoint, and then the formula $m(z) = f(0,z)$ works in all cases.

Proof. The formula for R from Theorem 3.12 shows that the convergence $R(L;z) \to 0$ of the radii of the Weyl disks is uniform on compact subsets of $\mathbb{C}^+$. Thus $m(z) = \lim_{L\to\infty} m_L^{(\beta)}(z)$ is holomorphic on $\mathbb{C}^+$ as a locally uniform limit of holomorphic functions.

In the second part, we can take $f = f_{m(z)} = T(m,1)^t$, which makes the asserted formula obvious. □

3.5 Spectral representation of canonical systems

We start with the problem on $[0, L]$. The important thing will be to have two regular endpoints, so everything I am going to say here also applies to $[0, \infty)$ if infinity is a regular endpoint. It is thus not necessary to pay further explicit attention to this case, and we will therefore assume that we have limit point case at infinity.

We always choose $L > 0$ so large that $(0, L)$ is not a single singular interval, so that all of our previous results apply.

So consider the self-adjoint realizations described by the boundary conditions

$$u_2(0) = 0, \quad u_1(L)\sin\beta - u_2(L)\cos\beta = 0.$$

By Theorem 3.4, the spectrum is purely discrete, and thus the normalized eigenfunctions form an orthonormal basis (ONB) of $\overline{D(S)}$ (but not necessarily of $L^2_H(0, L)$; we need the Hilbert space that the operator part S acts in). It is very easy to set up a spectral representation of such a relation: we send a vector to its expansion coefficients with respect to this ONB.

A suitable spectral measure is, very conveniently, delivered by the m function $m_L^{(\beta)}(z)$. For this, we need the *Herglotz representation theorem*, which says that if F is a Herglotz function, then there are unique $a \in \mathbb{R}$, $b \geq 0$, and a positive Borel measure ρ on $\mathbb{R}$ (possibly $\rho = 0$, but not both $b = 0, \rho = 0$) with $\int \frac{d\rho(t)}{1+t^2} < \infty$, such that

$$F(z) = a + bz + \int_{-\infty}^{\infty} \left(\frac{1}{t-z} - \frac{t}{t^2+1} \right) d\rho(t).$$

Conversely, for any a, b, ρ as above, this formula defines a Herglotz function.

The Herglotz representation may be viewed as a variant of the Poisson representation of positive harmonic functions, applied to $\operatorname{Im} F(z)$. For theoretical purposes, it is sometimes useful to rewrite it as

$$F(z) = a + \int_{\mathbb{R}_\infty} \frac{1+tz}{t-z}\, d\nu(t), \tag{3.9}$$

with $d\nu(t) = (1+t^2)^{-1}\, d\rho(t) + b\delta_\infty$. The main advantage of this version is that now ν is a finite measure (and $\nu \neq 0$) on the compact space $\mathbb{R}_\infty$.

So each Herglotz function comes with an associated measure, and one of the main points of the machinery built in the previous section is that the measure of the m function can serve as a spectral measure.

There are various results on how to extract the measure from F; for example,

$$d\rho(t) = \frac{1}{\pi}\, \mathrm{w}^*\!-\lim_{y\to 0+} \operatorname{Im} F(t+iy)\, dt, \tag{3.10}$$

and here the limit must be taken in weak-$*$ sense, so we will obtain convergence after integration against a compactly supported continuous function. Herglotz functions have normal limits $F(t) \equiv \lim F(t + iy)$ almost everywhere with respect to Lebesgue measure, and passing to this pointwise limit in (3.10) recovers the absolutely continuous part of ρ, that is, $d\rho_{ac}(t) = (1/\pi)\mathrm{Im}\, F(t)\, dt$. More importantly for us right now, the singular part of ρ is supported by the set where $\mathrm{Im}\, F(t) = \infty$, and specifically the point part corresponds to pole type behavior under normal approach. More precisely,

$$\rho(\{t\}) = \lim_{y\to 0+} -iyF(t + iy) \tag{3.11}$$

for all $t \in \mathbb{R}$.

The m function $m_L^{(\beta)}$ of our problem is meromorphic, and it is real on the real line, so ρ is a discrete measure with atoms precisely at the poles of m. We can be more specific: (3.2) shows that the poles occur precisely when the solution $f(x,z)$ that satisfies the boundary condition at $x = L$ also satisfies the one at $x = 0$, but this happens if and only if z is an eigenvalue. (Recall one more time the basic fact that if $(f, zf) \in \mathcal{S}$, then this element also goes into the operator part because an image in $\overline{D(\mathcal{S})}$ is exactly what we need, so z does become an eigenvalue of the operator S.)

We now compute $w \equiv \rho_L^{(\beta)}(\{t\})$ for such an eigenvalue $t \in \sigma(\mathcal{S}_L^{(\beta)})$. By (3.11) and (3.6) with equality (writing $m = m_L^{(\beta)}$),

$$\begin{aligned} w &= \lim_{y\to 0+} y\, \mathrm{Im}\, m(t + iy) = \lim_{y\to 0+} y^2 \int_0^L f_m^*(x)H(x)f_m(x)\, dx \\ &= \lim_{y\to 0+} y^2 |m(t + iy)|^2 \int_0^L u^*(x, t + iy)H(x)u(x, t + iy)\, dx \\ &= w^2 \int_0^L u^*(x,t)H(x)u(x,t)\, dx. \end{aligned}$$

To pass to the second line, I have used the fact that $ym(t+iy)$ stays bounded as $y \to 0+$, as does $\int p^*Hp$ for $p = u, v$. This calculation has shown that $w = 1/\|u(\cdot,t)\|^2$.

Theorem 3.16. *The measure $\rho = \rho_L^{(\beta)}$ that is associated with $m_L^{(\beta)}$ is given by*

$$\rho = \sum_{E\in\sigma(\mathcal{S}_L^{(\beta)})} \frac{\delta_E}{\|u(\cdot, E)\|^2},$$

and the map

$$U : L_H^2(0,L) \to L^2(\mathbb{R}, \rho),$$

$$(Uf)(t) = \int_0^L u^*(x,t)H(x)f(x)\, dx$$

sets up a spectral representation of $\mathcal{S}_L^{(\beta)}$.

Proof. We just derived the formula for ρ; note that this also holds if $\sigma(\mathcal{S}_L^\beta) = \emptyset$, which will happen if $(0, L)$ consists of two consecutive singular intervals, of types e_2 and $\beta^\perp$. In this case, $m(z) = a + bz$ and $\rho = 0$. (Theorem 3.19 below will provide some context for these remarks.)

The rest is then obvious from our earlier observation that the normalized eigenfunctions $u(\cdot, E_n)/\|u(\cdot, E_n)\|$, $\sigma(\mathcal{S}) = \{E_n\}$, form an ONB of $\overline{D(\mathcal{S})}$. Recall in this context that a self-adjoint operator with pure point spectrum $\{E_n\}$ and eigenvectors u_n, normalized so that they form an ONB, has domain $\{\sum a_n u_n : \sum |a_n|^2(1 + E_n^2) < \infty\}$, and of course it acts by multiplication by E_n on its eigenvector u_n. These remarks make it clear that $U\mathcal{S}U^*$ is indeed multiplication by t in $L^2(\mathbb{R}, \rho)$, on the correct domain. □

It could also be checked directly, by a computation, that $M_t Uf = Ug$ for $(f, g) \in \mathcal{S}$, as follows. Suppose also that $t \in \sigma(\mathcal{S}_L^{(\beta)})$. Then

$$t(Uf)(t) = t\int_0^L u^*(x,t)H(x)f(x)\,dx = \int_0^L u^{*\prime}(x,t)Jf_0(x)\,dx$$
$$= \int_0^L u^*(x,t)H(x)g(x)\,dx,$$

by an integration by parts. The boundary terms vanish because both $u(\cdot, t)$ and f_0 satisfy the boundary conditions. This step does not work in general if $t \notin \sigma(\mathcal{S}_L^{(\beta)})$, but these values are ignored by $\rho_L^{(\beta)}$, so are irrelevant in the Hilbert space $L^2(\mathbb{R}, \rho_L^{(\beta)})$.

While Theorem 3.16 is interesting and satisfying, its main role here is to serve as a stepping stone towards the analogous statement for the half line problem, which will take more work to establish. This half line problem has a unique m function $m(z)$ (recall that we assumed limit point case at infinity), and this $m(z)$ also comes with a measure ρ. From a formal point of view at least, everything then works the same way as before.

Theorem 3.17. *The recipe*

$$(Uf)(t) = \int_0^\infty u^*(x,t)H(x)f(x)\,dx, \quad f \in \bigcup_{L>0} L_H^2(0, L), \tag{3.12}$$
$$Uf = \lim_{L\to\infty} U(\chi_{(0,L)}f), \quad f \in L_H^2(0, \infty),$$

defines a map $U : L_H^2(0, \infty) \to L^2(\mathbb{R}, \rho)$; the limit will exist as a norm limit in $L^2(\mathbb{R}, \rho)$.

This map U together with the measure ρ provides a spectral representation of $\mathcal{S}$.

Note that $u(\cdot, t)$ will typically not lie in $L_H^2(0, \infty)$, so (3.12) certainly needs interpretation for a general $f \in L_H^2(0, \infty)$.

Proof. Our general strategy will be to obtain this from Theorem 3.16 by a limiting process $L \to \infty$. Recall that by general Weyl theory, $m_L^{(\beta)}(z) \to m(z)$ locally uniformly on $\mathbb{C}^+$, for any choice of boundary conditions $\beta = \beta(L)$. This implies (in fact, is essentially equivalent to) the weak-$*$ convergence of the associated measures from representation (3.9). This fact is well known and easy to establish; we will investigate these issues in a broader context (much) later, in Section 7.2, and the result that is relevant right now is formulated as Theorem 7.3 there.

If we state this convergence in terms of the measures ρ, then what we have is

$$\lim_{L\to\infty} \int f(t)\, d\rho_L^{(\beta)}(t) = \int f(t)\, d\rho(t) \tag{3.13}$$

for every continuous test function f with $\lim_{t\to\pm\infty} t^2 f(t) = 0$.

Consider now the relation

$$\mathcal{S}_0 = \{(f,g) \in \mathcal{S} : f_0(x) = 0 \text{ for all large } x\};$$

this is reminiscent of the relation $\mathcal{T}_{00}$ that we studied in Section 2.1, except that now f_0 is not required to be zero near $x = 0$. As before (compare Theorem 2.5(b)), we have $\overline{\mathcal{S}_0} = \mathcal{S}$. This can be derived formally from this result by observing that $\mathcal{S}_0 \supseteq \mathcal{T}_{00}$, so $\overline{\mathcal{S}_0} \supseteq \mathcal{T}_0$. On the other hand, $\overline{\mathcal{S}_0} \subseteq \mathcal{S}$. Since $\dim \mathcal{S} \ominus \mathcal{T}_0 = 1$, we see that $\overline{\mathcal{S}_0}$ can only be $\mathcal{T}_0$ or $\mathcal{S}$, and it is clearly not $\mathcal{T}_0$ because $\mathcal{S}_0$ contains elements (f,g) with $f_0(0) \neq 0$.

It will be convenient in the sequel to use the following notation: if $f \in L_H^2(0,\infty)$ is compactly supported, then we write

$$F(t) = \int_0^\infty u^*(x,t)H(x)f(x)\, dx.$$

This is a continuous function of t.

Our first goal is to show that if $(f,g) \in \mathcal{S}_0$, then

$$F \in L^2(\mathbb{R},\rho), \quad \|F\|_{L^2(\mathbb{R},\rho)} = \|f\|_{L_H^2(0,\infty)}. \tag{3.14}$$

The first claim is fairly obvious: if $f \in L_H^2(0,\infty)$ is compactly supported, then

$$\int_{-\infty}^\infty |F(t)|^2\, d\rho_L^{(\beta)}(t) \le \|f\|^2$$

for all large L, by Theorem 3.16. In particular, $\int \varphi_T |F|^2\, d\rho_L^{(\beta)} \le \|f\|^2$ for any continuous cut-off function $0 \le \varphi_T \le 1$, $\varphi_T(t) = 1$ for $|t| \le T$ and $\varphi_T(t) = 0$ for $|t| \ge T+1$, say. These integrals, however, approach $\int \varphi_T |F|^2\, d\rho$ as $L \to \infty$, by (3.13). Since $T > 0$ is arbitrary here, we obtain the desired conclusion that $F \in L^2(\mathbb{R},\rho)$, and in fact we have shown that

$$\|F\|_{L^2(\mathbb{R},\rho)} \le \|f\| \tag{3.15}$$

for any compactly supported $f \in L_H^2(0,\infty)$.

To go from here to (3.14) for $(f, g) \in \mathcal{S}_0$, we must show that no part of the finite measures $|F|^2\, d\rho_L^{(\beta)}$ can leak out to infinity when taking the limit $L \to \infty$, and this we do by bringing g into play also. By the calculation done after the proof of Theorem 3.16, $G(t) = tF(t)$, and this time, this is valid for all $t \in \mathbb{R}$; note that there are again no boundary terms from the integration by parts because f_0 is compactly supported and satisfies the boundary condition at $x = 0$, as does u.

By Theorem 3.16 and (3.15), the measures

$$(|F(t)|^2 + |G(t)|^2)\, d\rho_L^{(\beta)}(t), \quad (|F(t)|^2 + |G(t)|^2)\, d\rho(t)$$

are finite and uniformly bounded, each of them having total mass $\le \|f\|^2 + \|g\|^2$. Thus

$$\begin{aligned}\int_{-\infty}^{\infty} |F(t)|^2\, d\rho_L^{(\beta)}(t) &= \int_{-\infty}^{\infty} \frac{|F(t)|^2 + |G(t)|^2}{1+t^2}\, d\rho_L^{(\beta)}(t)\\ &= \int_{-\infty}^{\infty} \varphi_T(t) \frac{|F(t)|^2 + |G(t)|^2}{1+t^2}\, d\rho_L^{(\beta)}(t) + O(1/T^2)\end{aligned}$$

for any $T > 0$ and any cut-off function φ_T as above, and the constant implied in the error term can be taken to be independent of L. In the last integral, the integrand is now a compactly supported continuous function, so the integral converges to the corresponding integral with respect to $d\rho$. Since $T > 0$ is arbitrary here, it follows that

$$\lim_{L\to\infty} \int_{-\infty}^{\infty} |F(t)|^2\, d\rho_L^{(\beta)}(t) = \int_{-\infty}^{\infty} |F(t)|^2\, d\rho(t).$$

On the other hand, since $(f, g) \in \mathcal{S}_0$ also lies in $\mathcal{S}_L^{(\beta)}$ for all large L, the integrals on the left-hand side are all (eventually) equal to $\|f\|^2$, so (3.14) follows.

Having made substantial progress on the behavior of U on $\overline{D(\mathcal{S})}$, we now turn to its orthogonal complement $\mathcal{S}(0)$. I would now like to prove that if $g \in \mathcal{S}(0)$ is compactly supported, then $G(t) \equiv 0$, but this is not true in this form. It can fail when $(0, \infty)$ ends with a singular interval (c, ∞). This case is easy to analyze, but it requires a completely different argument and, therefore, I want to exclude it from the discussion for now. We will return to this at the very end of the proof.

With this extra assumption in place, the verification of my claim that $G(t) \equiv 0$ for $g \in \mathcal{S}(0) \cap L_H^2(0, L)$ poses no problems. Since we are now guaranteed the existence of regular points $> L$, we can conclude that the f_0 from $Jf_0' = -Hg$ is compactly supported, too. We then compute

$$G(t) = \int_0^{\infty} u^*(x,t) H(x) g(x)\, dx = -\int_0^{\infty} u^*(x,t) J f_0'(x)\, dx$$

$$= \int_0^\infty u^{*\prime}(x,t)Jf_0(x)\,dx = t\int_0^\infty u^*(x,t)H(x)f_0(x)\,dx = 0,$$

since $Hf_0 = 0$ almost everywhere.

Let us collect now what we currently have. We can define a continuous linear map

$$U_0 : D(\mathcal{S}_0) \oplus \left(\mathcal{S}(0) \cap \bigcup_{L>0} L^2_H(0,L) \right) \to L^2(\mathbb{R},\rho),$$

$$(U_0 f)(t) = \int_0^\infty u^*(x,t)H(x)f(x)\,dx,$$

and this is isometric on the first summand while it annihilates the second. Since the domain of U_0 is dense, this also gives us a unique continuous extension U to $L^2_H(0,\infty)$. This extended U is isometric on $\overline{D(\mathcal{S}_0)} = \overline{D(\mathcal{S})}$, and $N(U) = \mathcal{S}(0)$. It maps into $L^2(\mathbb{R},\rho)$.

For general $f \in L^2_H(0,\infty)$, it is currently not defined by (3.12), but by a slightly different limiting procedure. However, (3.12) will work, too, for the following reasons. Since U is continuous, it suffices to show that Uf satisfies (3.12) for $f \in L^2_H(0,L)$. Our construction defined $Uf := \lim F_n$, with $F_n = \int u^* H f_n$ and $f_n \in D(\mathcal{S}_0)\oplus(\mathcal{S}(0)\cap L^2_H(0,L_n))$, $f_n \to f$. Since f is itself compactly supported, we can also keep the supports of the f_ns inside a fixed compact set. Since u, restricted to this set, is in L^2_H, we have $F_n(t) \to \int u^*(x,t)H(x)f(x)\,dx$ pointwise, for any fixed $t \in \mathbb{R}$. On the other hand, the norm limit $F_n \to Uf$ becomes a pointwise limit ρ-almost everywhere after passing to a suitable subsequence, so the integral $\int u^* Hf$ does compute a representative of Uf, as claimed.

We already showed above that if $(f,g) \in \mathcal{S}_0$, then $Ug = tUf$, so, by taking suitable (norm) limits and making pointwise limits out of them on a subsequence, we see that also $(Ug)(t) = t(Uf)(t)$ for ρ-almost every t for general $(f,g) \in \mathcal{S}$. For any $g \in \overline{D(\mathcal{S})}$ and $z \notin \mathbb{R}$, we can let $f = (\mathcal{S}-z)^{-1}g \in D(\mathcal{S})$ and apply this observation to $(f, g+zf) \in \mathcal{S}$ to conclude that $(t-z)Uf = Ug$. This shows that whenever a function $G(t)$ lies in $R(U) \subseteq L^2(\mathbb{R},\rho)$, then so does $\frac{G(t)}{t-z}$. By repeating this step and taking linear combinations, we can produce a set of multipliers that is uniformly dense in $C(\mathbb{R}_\infty)$, by the Stone–Weierstraß theorem. Thus $hG \in R(U)$ if $G \in R(U)$ for any $h \in C(\mathbb{R}_\infty)$. This then implies the same property for $R(U)^\perp$, and from this we conclude that $R(U) = L^2(A,\rho)$, $R(U)^\perp = L^2(B,\rho)$, for some decomposition $\mathbb{R} = A \dot{\cup} B$. If we take a compactly supported $f \in L^2_H(0,\infty)$, then $F(t) \in R(U)$ is entire, so B can be taken as a discrete set. However, it is also an easy matter to make $F(t) \neq 0$ at any given $t \in \mathbb{R}$. We can return to Theorem 3.16 and use the fact that any $t \in \mathbb{R}$ will be in $\sigma(\mathcal{S}_L^{(\beta)})$ for some β. This follows because $u(L,t) \in \mathbb{R}^2$, so $u(\cdot,t)$ satisfies some boundary condition at $x = L$.

This whole argument has shown that $\rho(B) = 0$, and thus U, restricted to $\overline{D(\mathcal{S})}$, is unitary onto $L^2(\mathbb{R},\rho)$. Let us again denote this restriction by U_1. We already showed that $U_1(D(\mathcal{S})) \subseteq D(M_t)$ and $M_t U_1 f = U_1 \mathcal{S} f$, or $M_t F = U_1 \mathcal{S} U_1^* F$ (writing $F = U_1 f$) for $f \in D(\mathcal{S})$. This says that the self-adjoint operator $U_1 \mathcal{S} U_1^*$ is a restriction of M_t, but M_t is self-adjoint itself, so $M_t = U_1 \mathcal{S} U_1^*$.

The proof is complete, except for the postponed case of a singular half line (c,∞), which we now tackle. Let α be the type of (c,∞). We will show that then we are effectively dealing with the problem on $[0,c]$, with boundary condition $\alpha^\perp$ at $x=c$.

Start out by observing that Theorem 2.27 shows that $L^2_H(c,\infty)\subseteq\mathcal{S}(0)$, so indeed (c,∞) disappears from the Hilbert space when we pass to $\overline{D(\mathcal{S})}$. If $(f,g)\in\mathcal{S}$, then $P_\alpha f_0(x)$ is constant on (c,∞), and since $H\notin L^1(c,\infty)$, it follows that $P_\alpha f_0(c)=0$, that is, f_0 satisfies the boundary condition $\alpha^\perp$ at $x=c$. Conversely, if $(f,g)\in\mathcal{S}_c^{(\alpha^\perp)}$, then $P_\alpha f_0(c)=0$, so if we extend by setting $f_0(x)=f_0(c)$, $g(x)=0$ on (c,∞), then the extended pair will represent an element of $\mathcal{S}$. We have shown that

$$\mathcal{S}=\mathcal{S}_c^{(\alpha^\perp)}\oplus(0,L^2_H(c,\infty)).$$

Now let us take a look at the m function, which was defined as $m(z)=f(0,z)$, with $Jf'=-zHf$, $f\in L^2_H(0,\infty)$. By the argument we just gave, f will be in L^2_H if and only if it satisfies the boundary condition $\alpha^\perp$ at $x=c$. Hence also $m(z)=m_c^{(\alpha^\perp)}(z)$.

We are back in the situation of Theorem 3.16, and everything is clear now, except for the detail that originally gave us pause about this special situation: what does U from Theorem 3.17 do to the $f\in L^2_H(c,\infty)\subseteq\mathcal{S}(0)$? This is also easy to clarify now. If $t\in\sigma(\mathcal{S})=\sigma(\mathcal{S}_c^{(\alpha^\perp)})$, then $P_\alpha u(c,t)=0$, so $H(x)u(x,t)=0$ on (c,∞). Hence Uf does represent the zero element of $L^2(\mathbb{R},\rho)$, as required. □

This special situation analyzed in the final part of the proof is sufficiently important to deserve a separate summary.

Theorem 3.18. *Suppose that (L,∞) is a singular interval of type $\beta^\perp$, with limit point case at infinity. Then $L^2_H(L,\infty)\subseteq\mathcal{S}(0)$, $\mathcal{S}=\mathcal{S}_L^{(\beta)}\oplus(0,L^2_H(L,\infty))$, $S=S_L^{(\beta)}$, and $m(z)=m_L^{(\beta)}(z)$.*

Or, in short, a singular half line with $H\notin L^1$ there implements a boundary condition at its finite endpoint.

In the very special (and trivial) situation when a singular interval $(0,L)$ is immediately followed by a singular half line, Theorem 3.18 says that we effectively obtain the problem on $(0,L)$, which consists of a single singular interval, and we seem to have outmaneuvered ourselves. That is not the case, as we can easily confirm by working everything out explicitly. For example, in the most extreme case, when the type of $(0,L)$ is e_2, we obtain $m(z)=a+bz$, with $b=\int_0^L h(x)\,dx$, and a depends on the type of (L,∞) (this was already pointed out, in a slightly different context, at the beginning of the proof of Theorem 3.16). So $\rho=0$, which is consistent with the fact that $\mathcal{S}(0)=L^2_H(0,\infty)$. If the type of $(0,L)$ is distinct from e_2, then ρ will be supported by the single point in $\sigma(\mathcal{S})$, and again everything is in perfect order.

In this whole analysis, the spaces $\overline{D(\mathcal{S})}=\mathcal{S}(0)^\perp$ played a central role. We of course have a precise description of them from the material of Section 2.4, and, given their importance, it seems useful to make this completely explicit. We know that we have to distinguish cases if our basic interval starts or ends with a singular interval. It will be convenient to treat together the two cases of a bounded interval and a half line, so we will write the basic interval as $(0, L)$, and now $0 < L \le \infty$. As usual, we assume limit point case at L if $L = \infty$, and if $L < \infty$, then we impose the boundary condition β at L. The corresponding self-adjoint relation will be denoted by $\mathcal{S}$ in all cases. Finally, we denote the set of regular points by $R \subseteq (0, L)$.

Theorem 3.19. *We have*

$$\overline{D(\mathcal{S})} = \mathcal{S}(0)^{\perp} = L^2_H(R) \oplus \bigoplus_j L(\chi_{(c_j,d_j)} e_{\alpha_j})$$

(the sum is over the singular intervals $(c_j, d_j) \subseteq (0, L)$, and their types have been denoted by α_j), except for the following possible modifications in the case of initial or final singular intervals: if $(0, L)$ starts with a singular interval of type e_2 or ends with a singular interval of type $\beta^{\perp}$ or $L = \infty$ and $(0, \infty)$ ends with a singular half line, then the summand corresponding to such an interval is absent.

Proof. This just rephrases the description of $\mathcal{S}(0)$ that was given in Theorems 2.27, 2.28. □

This looks a bit unwieldy due to the various cases that can arise, but of course the statement has an easy summary: each singular interval contributes a one-dimensional subspace to $\mathcal{S}(0)^{\perp}$, containing the constant (on that interval) function e_{α_j}, and if an initial or final singular interval is of a certain exceptional type, then it will not contribute anything.

Moreover, Theorem 3.18 provides a clear intuitive explanation why these exceptional types of initial or final singular intervals are special. Consider, for example, a canonical system on $(0, L)$ with boundary condition β at $x = L$ and a singular interval (c, L) of type $\beta^{\perp}$. We can then imagine the boundary condition being implemented by a singular half line (L, ∞) of type $\beta^{\perp}$, and then (c, L) is no longer a separate singular interval; rather, it just prolongs this singular half line. The same remarks apply to an initial singular interval of type e_2.

3.6 General boundary conditions

The whole theory of course works in the same way for a general boundary condition

$$u_1(0) \sin \alpha - u_2(0) \cos \alpha = 0$$

(instead of specifically $\alpha = 0$) at $x = 0$. Among our many characterizations of m, the one that generalizes most naturally is the condition that $v + mu$ lies in $D(\mathcal{S})$ near the right endpoint. The solutions u, v are what we need to adapt now. In the original setting, we had $u(0, z) = e_1$, $v(0, z) = e_2$, so the value of $u(0)$ is the unit vector satisfying the boundary condition, the value of $v(0)$ is the orthogonal unit vector, and $W(v, u) = 1$. Taking this as our cue, we are now led to define two new solutions by putting $u_\alpha(0) = e_\alpha$, $v_\alpha(0) = e_{\alpha^\perp}$, and then we define $m = m_\alpha$ by demanding that $f = v_\alpha + m_\alpha u_\alpha$ lies in the domain near the right endpoint. More explicitly, f is required to satisfy the boundary condition at $x = L$ if we are dealing with the problem on $[0, L]$, and if we are on $[0, \infty)$, with limit point case at infinity, then we ask that $f \in L^2_H(0, \infty)$.

As usual, transfer matrices can handle all this more gracefully. The solutions u_α, v_α are the columns of

$$T_\alpha(x; z) \equiv T(x; z) R_\alpha, \quad R_\alpha = \begin{pmatrix} \cos\alpha & -\sin\alpha \\ \sin\alpha & \cos\alpha \end{pmatrix},$$

and then m_α may be described as the unique number with the property that

$$T_\alpha(x; z) \begin{pmatrix} m_\alpha \\ 1 \end{pmatrix} = T(x; z) R_\alpha \begin{pmatrix} m_\alpha \\ 1 \end{pmatrix}$$

lies in the domain near the right endpoint. However, the number that works for T is m, so

$$m_\alpha(z) = R_{-\alpha} m(z). \tag{3.16}$$

It would be easy now to restore α in the discussion of Section 3.4 and give a general treatment, and the reader can do this if interested, but it really seems quite unnecessary, thanks to the following result.

Theorem 3.20. *Let $A \in \mathrm{SL}(2, \mathbb{R})$, put $H_A(x) = A^{-1t} H(x) A^{-1}$, and consider the canonical systems H and H_A on $[0, \infty)$, with limit point case at infinity and boundary condition $u_2(0) = 0$. Then $m_A(z) = Am(z)$.*

So, comparing this with (3.16), we see that a change of boundary condition at $x = 0$ can be simulated by transforming H instead, using the rotation matrix R_α, while keeping the boundary condition the same. By Theorem 3.18, problems on $[0, L]$ are covered by this; note, however, that the boundary condition at $x = L$ now gets transformed also. We have $Am^{(\beta)} = m_A^{(\beta_A)}$, with $\cot\beta_A = A\cot\beta$. If specifically $A = R_{-\alpha}$, then $\beta_A = \beta - \alpha$.

Proof. By Theorem 1.6, $T_A = ATA^{-1}$, so

$$T_A(x) \begin{pmatrix} Am \\ 1 \end{pmatrix} = cT_A(x) A \begin{pmatrix} m \\ 1 \end{pmatrix} = cAT(x) \begin{pmatrix} m \\ 1 \end{pmatrix},$$

and if we compute the $L^2_{H_A}$ norm of this solution, then the extra factor of A on the left cancels, so this is the $L^2_{H_A}$ solution, as claimed. □

3.7 Two limit point endpoints

We now discuss the spectral representation of a canonical system with two limit point endpoints. It will be convenient to assume that the basic interval is $\mathbb{R}$ itself. We will cut the real line into the two half lines $(-\infty, 0)$ and $(0, \infty)$ and then use the spectral data (especially m functions) of these half line problems to set up a spectral representation of the original problem also.

In this abstract description, this does not sound like a particularly natural or promising idea, and it also breaks the symmetry by making $x = 0$ a distinguished point, but actually this is a standard method that is in common use for other operators such as Schrödinger or Jacobi operators. Its great advantage is that we may still work with m functions throughout even though there is no direct definition of these for the whole line problem.

We will then need the theory of Sections 3.4, 3.5 also for left half lines $(-\infty, 0)$, where now the right endpoint is the regular endpoint. Of course, this presents no problems whatsoever. The only (very minor) change occurs right at the beginning, in the definition of the m function of this problem, with boundary condition $u_2(0) = 0$, as always: it is now given by $m_-(z) = -f(0, z)$, $z \in \mathbb{C}^+$, and here f denotes a non-trivial solution of $Jf' = -zHf$ with $f \in L^2_H(-\infty, 0)$. It is perhaps worth pointing out that if you want to reduce matters directly to a right half line problem by reflecting, then m_- becomes the m function not of $H(-x)$, but of $H_1(x) = IH(-x)I$, with $I = \text{diag}(1, -1)$, $x > 0$.

Of course, we now also have a second m function m_- available for the problems on bounded intervals $(L, 0)$, in addition to the one discussed in Section 3.4, if we let the endpoints swap roles. It is also given by $m_-(z) = -f(0, z)$, and now f is a solution that satisfies the boundary condition at $x = L$.

Everything works the same way as before. In particular, these new m functions $m_-(z)$ are also Herglotz functions.

The key object in the spectral representation of the whole line relations will be the matrix valued function

$$M(z) = \frac{-1}{2(m_+ + m_-)} \begin{pmatrix} -2m_+m_- & m_+ - m_- \\ m_+ - m_- & 2 \end{pmatrix}; \qquad (3.17)$$

here, m_+ and m_- denote the half line m functions of the problems on $(0, \infty)$ and $(-\infty, 0)$, respectively, and of course all functions are evaluated at z. This function M is a *matrix valued Herglotz function*, that is, $M(z)$ is holomorphic on $z \in \mathbb{C}^+$, and $\text{Im}\, M(z) > 0$ there, with $\text{Im}\, M = (1/2i)(M - M^*)$, and the inequality is now understood as saying that $\text{Im}\, M$ is a strictly positive definite matrix. Our map $M(z)$ has the additional property that $M = M^t$, so maps into what is called *Siegel upper half space*. It also satisfies $\det M = -1/4$.

Let us now verify this claim I just made, that $\text{Im}\, M(z) > 0$. Clearly, the diagonal elements of this matrix are positive because they are the imaginary parts of the Herglotz functions $-1/(m_+ + m_-)$ and

$$\frac{m_+ m_-}{m_+ + m_-} = \frac{-1}{\frac{-1}{m_+} + \frac{-1}{m_-}}.$$

So we can now confirm the asserted positivity by looking at the determinant, and a calculation shows that

$$\det \operatorname{Im} M(z) = \frac{\operatorname{Im} m_+(z) \operatorname{Im} m_-(z)}{|m_+(z) + m_-(z)|^2},$$

which is positive, as required.

The Herglotz representation theorem generalizes to matrix valued Herglotz functions without any problems. One method would be to look at the (scalar) Herglotz functions $v^* M(z) v$, for $v \in \mathbb{C}^2$, and treat the off-diagonal functions by polarization. Thus, in our setting, there are $A, B \in \mathbb{R}^{2\times 2}$, $B \geq 0$, and a matrix valued Borel measure ρ on $\mathbb{R}$ such that

$$M(z) = A + Bz + \int_{-\infty}^{\infty} \left(\frac{1}{t-z} - \frac{t}{t^2+1} \right) d\rho(t).$$

The measure ρ assigns a positive definite matrix $\rho(S)$ to a bounded Borel set $S \subseteq \mathbb{R}$. Since $M = M^t$, we will have $\rho(S) \in \mathbb{R}^{2\times 2}$.

This measure ρ will again serve as the spectral measure. Since it is matrix valued, we are now in a situation that was not foreseen in Definition 3.1 in exactly this form, so we also need to adapt the notion of a spectral representation: what we will construct will be a partial isometry $U : L^2_H(\mathbb{R}) \to L^2(\mathbb{R}, \rho)$ with $N(U) = \mathcal{S}(0)$ that maps $\overline{D(\mathcal{S})}$ unitarily onto $L^2(\mathbb{R}, \rho)$, such that $U\mathcal{S}U^* = M_t$ is again the self-adjoint operator of multiplication by the variable in $L^2(\mathbb{R}, \rho)$.

This space, $L^2(\mathbb{R}, \rho)$, formed with a matrix valued measure, is of course defined in the same way as L^2_H: its elements are represented by functions $F : \mathbb{R} \to \mathbb{C}^2$, and norms are computed as follows:

$$\|F\|^2 = \int_{-\infty}^{\infty} F^*(t)\, d\rho(t) F(t)$$

(in this order). It is the same type of space as L^2_H, except that now the measure is not assumed to be absolutely continuous. One could also introduce $\mu(S) = \operatorname{tr} \rho(S)$ and then write $d\rho(t) = D(t)\, d\mu(t)$ as a matrix valued density times a scalar measure, so matrix valued measures, while convenient, are not really needed here.

Let us now try to use this machinery to set up a spectral representation (in this more general sense) of $\mathcal{S}$, the self-adjoint realization of our canonical system, which agrees with both the minimal and maximal relations, by the limit point assumptions on the endpoints. As our overall strategy, we will follow the treatment of Section 3.5 rather closely. In particular, we again start with the analysis of the truncated problems on (L_-, L_+), with $L_- < 0 < L_+$.

As always, we make our basic assumption that (L_-, L_+) is not just a single singular interval. This does not prevent $(L_-, 0)$ or $(0, L_+)$ from being contained in a single singular interval, but let us for now proceed under the additional assumption that this is not the case. Later, we will actually need those cases also, but let us first see the whole argument without being distracted by too many technical details.

We impose boundary conditions $\beta_\pm$ at $L_\pm$ and then also have m functions $m_\pm$ of these problems on $(L_-, 0)$ and $(0, L_+)$ available. I will use this simplified notation $m_\pm(z)$ in this part of the analysis, but keep in mind that each m also depends on the choice of the corresponding L and the boundary condition there.

We then use these $m_\pm$ to define a matrix valued Herglotz function $M(z)$ by (3.17). This function is rational and real on the real line, so ρ is a discrete measure. We could analyze it directly, more or less in the same way as in Section 3.5. This would require some additional effort, though, and it is technically more convenient to use a different approach, which will relate M to the resolvent $(\mathcal{S} - z)^{-1}$ of this problem on (L_-, L_+). This has the added advantage that it will provide some motivation for (3.17), which did not seem very intuitive originally.

We return to Theorem 3.3. We then need two solutions, one for each endpoint, that satisfy the boundary conditions, and we have two natural candidates, namely

$$f_\pm(x, z) = T(x, z) \begin{pmatrix} \pm m_\pm(z) \\ 1 \end{pmatrix}. \tag{3.18}$$

Their Wronskian can be conveniently evaluated at $x = 0$, and we find

$$W(f_-, f_+) = (-m_-(z), 1) J \begin{pmatrix} m_+(z) \\ 1 \end{pmatrix} = m_+(z) + m_-(z);$$

so if we remember to also divide by $-(m_+ + m_-)$, then these solutions will work as u_a, u_b in Theorem 3.3. So if, for example, $L_- < x \le y < L_+$, then the Green function is given by

$$\begin{aligned} G(x, y; z) &= \frac{-1}{m_+ + m_-} f_-(x, z) f_+^t(y, z) \\ &= \frac{-1}{m_+ + m_-} T(x; z) \begin{pmatrix} -m_+ m_- & -m_- \\ m_+ & 1 \end{pmatrix} T^t(y; z), \end{aligned}$$

and of course there is a similar formula which is valid for $x > y$, and there the matrix in the middle will be the transpose of the one we just found. We can combine these two formulae by writing

$$G(x, y; z) = T(x; z) M(z) T^t(y; z) \pm \frac{1}{2} T(x; z) J T^t(y; z);$$

the choice of sign is determined by whether $x \le y$ or $x > y$. This does already add some context to (3.17), as promised; note, however, that it was not mandatory to introduce M exactly by the formula given. We can also write

$$M(z) = \frac{-1}{m_+ + m_-}\begin{pmatrix} -m_+m_- & m_+ \\ m_+ & 1 \end{pmatrix} + \frac{1}{2}\begin{pmatrix} 0 & 1 \\ 1 & 0 \end{pmatrix}$$
$$= \frac{-1}{m_+ + m_-}\begin{pmatrix} -m_+m_- & -m_- \\ -m_- & 1 \end{pmatrix} - \frac{1}{2}\begin{pmatrix} 0 & 1 \\ 1 & 0 \end{pmatrix},$$

and either of these two related matrix functions would have worked too as our basic object. Note that all three functions have the same imaginary part.

After this brief digression, let us now return to the analysis of the resolvent. Take an $f \in L^2_H(L_-, L_+)$. We can then write

$$\langle f, (\mathcal{S} - z)^{-1} f\rangle = \int_{L_-}^{L_+} dx \int_{L_-}^{L_+} ds\, f^*(x)H(x)G(x,s;z)H(s)f(s)$$
$$= \iint dx\, ds\, f^*(x)H(x)T(x;z)M(z)T^*(s;\overline{z})H(s)f(s)$$
$$+ \iint dx\, ds\, f^*(x)H(x)T(x;z)\sigma J T^*(s;\overline{z})H(s)f(s), \qquad (3.19)$$

with $\sigma = \sigma(x,s) = \pm 1/2$.

Recall that $(\mathcal{S} - z)^{-1}$ annihilates $\mathcal{S}(0)$ and maps into $\overline{D(\mathcal{S})} = \mathcal{S}(0)^\perp$ (Theorem 3.1). So f can be replaced by its projection $f_0 = Pf$ onto $\overline{D(\mathcal{S})}$. By functional calculus, if E denotes the spectral resolution of the self-adjoint operator S on $\overline{D(\mathcal{S})}$, then

$$\langle f_0, (S - z)^{-1} f_0\rangle = \int_{-\infty}^{\infty} \frac{d\|E(t)f_0\|^2}{t - z}.$$

In particular, this is a Herglotz function, with associated measure $d\mu(t) = d\|E(t)f_0\|^2$, and since $E(\{t\})$ projects onto the space of eigenvectors with eigenvalue t, which is one-dimensional if t actually is an eigenvalue, we have

$$\mu(\{t\}) = |\langle \varphi_t, f_0\rangle|^2 = |\langle \varphi_t, f\rangle|^2$$

for such a t, with φ_t denoting a normalized eigenfunction. It follows that

$$\lim_{y\to 0+} -iy\langle f, (\mathcal{S} - t - iy)^{-1} f\rangle = \begin{cases} |\langle \varphi_t, f\rangle|^2, & t \in \sigma(\mathcal{S}), \\ 0, & \text{otherwise.} \end{cases}$$

We now compute this limit directly, using (3.19). Notice, first of all, that the term with $\pm J$, from the last line, does not contribute to this because $T(x; t + iy)$ is continuous on $L_- \le x \le L_+$, $y \ge 0$, and $Hf \in L^1(L_-, L_+)$. For the same reason, we can then take the limit of $-iyM(t+iy)$ separately and simply replace $t+iy$ by t in $T(x; t+iy)$, $T^*(s; t-iy)$. We then have $\lim_{y\to 0+} -iyM(t+iy) = \rho(\{t\})$, by using the same general fact about Herglotz functions that we just referred to (tacitly) a moment ago. Putting everything together, we thus see that

$$F^*(t)\rho(\{t\})F(t) = \begin{cases} |\langle \varphi_t, f\rangle|^2, & t \in \sigma(\mathcal{S}), \\ 0, & \text{otherwise}, \end{cases} \tag{3.20}$$

$$F(t) \equiv \int_{L_-}^{L_+} T^*(x; t)H(x)f(x)\,dx.$$

The relation $\mathcal{S}$ on (L_-, L_+) has purely discrete simple spectrum, so the eigenfunctions $\{\varphi_t : t \in \sigma(\mathcal{S})\}$ form an ONB of $\overline{D(\mathcal{S})}$. So, if we now introduce, for $f \in L^2_H(L_-, L_+)$,

$$(Uf)(t) = \int_{L_-}^{L_+} T^*(x; t)H(x)f(x)\,dx,$$

then what (3.20) is saying is that U maps $\overline{D(\mathcal{S})}$ isometrically into $L^2(\mathbb{R}, \rho)$, and $N(U) = \mathcal{S}(0)$. In fact, U maps $\overline{D(\mathcal{S})}$ unitarily onto $L^2(\mathbb{R}, \rho)$, and this follows because if $t \in \mathbb{R}$, $v \in \mathbb{C}^2$, $v \ne 0$, are given, then there will be an $f \in L^2_H(L_-, L_+)$ such that $v^*F(t) \ne 0$. To confirm this latter claim, suppose that, on the contrary, we had

$$v^* \int_{L_-}^{L_+} T^*(x; t)H(x)f(x)\,dx = 0$$

for all $f \in L^2_H(L_-, L_+)$. This would then imply that $H(x)T(x; t)v = 0$ for almost all $x \in (L_-, L_+)$, but this makes $T(x; t)v$ a constant function (of x) and thus (L_-, L_+) a singular interval.

Now let us take a closer look at $\rho(\{t\})$ for a $t \in \sigma(\mathcal{S})$. For such a t, the solution f_- from (3.18) that satisfies the boundary condition at $x = L_-$ must be a multiple of the solution f_+ that satisfies the other boundary condition, at $x = L_+$. These solutions are well defined for $t \in \mathbb{R}$ also since $m_\pm$ are meromorphic, except that it could happen that m_- or $m_+ = \infty$. If, say, $m_-(t) = \infty$, then $Te_1 = u$ is the solution satisfying the boundary condition at L_-.

All this means that $t \in \sigma(\mathcal{S})$ if and only if either $m_+(t) = -m_-(t)$, or $m_+(t) = m_-(t) = \infty$. Let us discuss the first case, when $m_+(t) = -m_-(t) \in \mathbb{R}$, in more detail; the second one is similar. Then $-1/(m_+ + m_-)$ has a pole at $z = t$, while the matrix from the right-hand side of (3.17) is holomorphic near $z = t$. Thus

$$\rho(\{t\}) = \lim_{y\to 0+} -iyM(t+iy) = w\begin{pmatrix} m_+(t)^2 & m_+(t) \\ m_+(t) & 1 \end{pmatrix} \tag{3.21}$$

for some $w > 0$. Notice that this matrix is singular, as it must be because $\mathcal{S}$ has only simple eigenvalues.

We now finish our discussion of the problem on (L_-, L_+), from the point of view of the matrix valued Herglotz function $M(z)$, by showing that if $(f, g) \in \mathcal{S}$, then $g(t) = t(Uf)(t)$ almost everywhere with respect to ρ. This will then imply, by the arguments familiar to us from Section 3.5, that $U\mathcal{S}U^* = M_t$, and thus the map U together with the measure ρ gives us a spectral representation of $\mathcal{S}$.

So let $(f, g) \in \mathcal{S}$. As usual, we compute

$$\begin{aligned}(Ug)(t) &= \int_{L_-}^{L_+} T^*(x;t)H(x)g(x)\,dx = -\int_{L_-}^{L_+} T^*(x;t)Jf_0'(x)\,dx \\ &= T^*(L_-;t)Jf_0(L_-) - T^*(L_+;t)Jf_0(L_+) \\ &\quad + t\int_{L_-}^{L_+} T^*(x;t)H(x)f_0(x)\,dx, \end{aligned} \tag{3.22}$$

by an integration by parts. The last term equals $t(Uf)(t)$, so we must show that the boundary terms are zero. We know that f_0 satisfies the boundary conditions, so the first term for example is a multiple of $T^*(L_-;t)Je_{\beta_-}$. If now also $t \in \sigma(\mathcal{S})$, then f_+ satisfies both boundary conditions, so $e_{\beta_-}^* Jf_+(L_-, t) = 0$, or, equivalently,

$$(m_+(t), 1)T^*(L_-;t)Je_{\beta_-} = 0.$$

Now compare this with (3.21): we have shown that $\rho(\{t\})$ annihilates the first term from (3.22) for all $t \in \sigma(\mathcal{S})$. More precisely, we have explicitly shown this only in the case $m_\pm(t) \in \mathbb{R}$, but of course a similar argument works in the other case, too. The boundary term at L_+ also represents the zero element of $L^2(\mathbb{R}, \rho)$, for the same reasons, so $Ug = M_t Uf$, as claimed.

Before we can proceed to the analysis of the whole line problem, we must remove the extra assumption that we made at the beginning that neither $(L_-, 0)$ nor $(0, L_+)$ is contained in a single singular interval. We do still assume, as always, that (L_-, L_+) itself is not a single singular interval.

In fact, before we can even begin, we must generalize our definition of the m function to this setting. This, fortunately, presents no problems. The formulae $m_\pm(z) = \pm f_\pm(0, z)$, with $f_\pm$ satisfying the boundary conditions, work in all cases; they now define generalized Herglotz functions. The formula defining M makes sense in this case also, if suitably interpreted in case m_+ or $m_- \equiv \infty$; note here that even if one or both of $m_\pm$ are constant, we cannot have $m_+ + m_- \equiv 0$ or $m_- = m_+ \equiv \infty$ since that would make (L_-, L_+) a single singular interval. A difference with the previous case is that this time, we can only say that $\operatorname{Im} M(z) \geq 0$ (not strictly positive) for $z \in \mathbb{C}^+$.

Now, if we go through the whole argument one more time, we see that actually almost nothing changes in this more general setting. The analysis of the resolvent that established the mapping properties of U works in general, if we remember to replace the corresponding f by u in case an m function is identically equal to infinity. I will discuss similar issues below, so I leave the details of this straightforward adjustment to the reader.

The calculation leading to (3.22) did not even mention m functions, so remains valid. The only part of the analysis that needs to be revisited is formula (3.21), but again, it is easy to confirm that this still works if m_+ or $m_- \equiv a \in \mathbb{R}_\infty$.

Let me perhaps discuss one rather degenerate scenario in detail, to add further detail to these remarks and also have the case covered that we skipped in (3.21).

Assume now that (L_-, d) is a singular interval of type specifically e_2, for some $d \geq 0$, and let us further assume that $e_{\beta_-} = e_1$. Then $m_-(z) = \infty$, and this gives us

$$M(z) = \begin{pmatrix} m_+(z) & 1/2 \\ 1/2 & 0 \end{pmatrix}.$$

In particular, the associated measure is simply given by

$$d\rho(t) = \begin{pmatrix} 1 & 0 \\ 0 & 0 \end{pmatrix} d\rho_+(t).$$

Now take a look at (3.22). The first boundary term $T^*(L_-;t)Jf_0(L_-)$ now becomes a multiple of $T^*(L_-;t)e_2$, but $e_2^* T(x)$ is constant on (L_-, d), so this equals $T^*(0;t)e_2 = e_2$ and is thus annihilated by ρ, as required. Similarly, the other boundary term will vanish if we can show that $e_{\beta_+}^* JT(L_+;t) = 0$ in $L^2(\mathbb{R}, \rho)$, and this follows exactly as before because a t that is given positive weight by ρ_+ is a pole of m_+, so the solution Te_1 satisfies the boundary condition β_+ at such a t. Hence $e_{\beta_+}^* JT(L_+;t)e_1 = 0$, and this says that $\rho(\{t\})$ annihilates the boundary term.

We are now finally ready for the analysis of the whole line problem. We will prove the following analog of Theorem 3.17.

Theorem 3.21. *The recipe*

$$(Uf)(t) = \int_{-\infty}^{\infty} T^*(x,t)H(x)f(x)\,dx, \quad f \in \bigcup_{L>0} L_H^2(-L, L),$$
$$Uf = \lim_{L\to\infty} U(\chi_{(-L,L)}f), \quad f \in L_H^2(\mathbb{R}),$$

defines a map $U : L_H^2(\mathbb{R}) \to L^2(\mathbb{R}, \rho)$; the limit will exist as a norm limit in $L^2(\mathbb{R}, \rho)$.

This map U together with the measure ρ provides a spectral representation of $\mathcal{S}$.

Singular half lines are allowed here, and, as before, they implement boundary conditions. In particular, $(-\infty, 0)$ or $(0, \infty)$ itself or even both of these could be a singular half line, and thus, similarly to what we just discussed, we will now also need half line m functions for these coefficient functions. Again, this presents no problems. We still define $m_\pm(z) = \pm f_\pm(0,z)$ as the value at $x = 0$ of the solution $f_\pm$ that is in L^2_H on the corresponding half line. If, say, $(0,\infty)$ is a singular half line of type α, then we can take $f(x,z) = e_{\alpha^\perp}$, so $m_+(z) = -\tan\alpha$. In particular, if $\alpha = \pi/2$ here, then $m_+(z) = \infty$. Singular half lines give us exactly the (trivial) generalized Herglotz functions $m \equiv a \in \mathbb{R}_\infty$.

Also as above, we can then still define M by (3.17), with the obvious interpretation of this formula applied if m_+ or $m_- = \infty$. If both half lines are singular, then M is simply a constant real matrix, corresponding to the fact that $\mathcal{S} = (0, L^2_H(\mathbb{R}))$ does not have an operator part, and thus $\sigma(\mathcal{S}) = \emptyset$. We cannot have $m_+ = -m_-$ or both $m_\pm = \infty$ in this trivial scenario because this would give both singular half lines the same type, and we are still excluding this scenario.

Proof. The proof of Theorem 3.17 will provide a template for this, which we will follow very closely. In my presentation here, I will be brief on those parts that almost literally repeat the arguments from that proof, and I will focus on the new aspects, of which there actually will not be too many.

In the first part of the argument, we work with the relation $\mathcal{T}_{00}$ (the notation is from Chapter 2) that contains those $(f,g) \in \mathcal{S} = \mathcal{T}$ for which f_0 is compactly supported. We again use the notation

$$F(t) = \int_{-\infty}^{\infty} T^*(x;t)H(x)f(x)\,dx$$

for compactly supported $f \in L^2_H(\mathbb{R})$. In the same way as before, we can then show that $F \in L^2(\mathbb{R},\rho)$ for any such f and $\|F\| \le \|f\|$. Moreover, if $(f,g) \in \mathcal{T}_{00}$, then $\|F\| = \|f\|$.

Now we again distinguish cases. Let us first assume that there are no singular half lines. Then the same calculation as before shows that if $g \in \mathcal{S}(0)$ is compactly supported, then $G(t) \equiv 0$. We are now in a position to establish that a map U with the desired mapping properties can indeed be defined in the intended way, having already defined it on a dense subspace.

By its construction, U maps $\overline{D(\mathcal{S})}$ isometrically into $L^2(\mathbb{R},\rho)$, and we must now show that it in fact maps *onto* this space. We start the argument as before, and we see in the same way that if $G \in R(U)$, then also $hG \in R(U)$ for any $h \in C(\mathbb{R}_\infty)$, and $R(U)^\perp$ then has the same property. We also see as above that there are $G \in R(U)$ that do not vanish on any set of positive ρ measure, but unlike in the situation of Theorem 3.17, that does not immediately finish the proof of the surjectivity of U because functions in $L^2(\mathbb{R},\rho)$ now have two components. However, now we can make use of the following fact, which was already established above: for any $t \in \mathbb{R}$ and $v \in \mathbb{C}^2$, $v \ne 0$, there is a compactly supported $f \in L^2_H(\mathbb{R})$ such that $v^*F(t) \ne 0$ (recall that F is continuous, so this will also hold in a neighborhood of t then). If this is combined with what we already have, then it follows that $R(U) = L^2(\mathbb{R},\rho)$.

Finally, it is then also easy to see that $USU^* = M_t$, by first verifying that $G = tF$ for $(f,g) \in \mathcal{T}_{00}$ and then arguing as in the proof of Theorem 3.17.

It remains to discuss the case when we have one or two singular half lines. To make this more specific, let us assume that (L, ∞) is a singular interval, of type α; here, $L \in \mathbb{R}$ could have any value and, in particular, $L \leq 0$ is possible.

This case is not fundamentally different from what we did before, and the only issue that needs additional attention is that of the transform G of a $g \in \mathcal{S}(0)$ that is supported by (L, ∞). We must show that still $G(t) = 0$ in this situation, and the new aspect is that this will now not hold for all $t \in \mathbb{R}$, but only as an element of $L^2(\mathbb{R}, \rho)$, almost everywhere with respect to ρ.

To do this, we reintroduce $L_- < 0$, which we will send to $L_- \to -\infty$ eventually, and consider the problem on (L_-, ∞), with some boundary condition at $x = L_-$. We form the matrix M with this modified $m_- = m_{-,L_-}^{(\beta_-)}$ and the m_+ we actually have. If $L > 0$ here, then m_+ becomes the m function of the problem on $(0, L)$, with the boundary $\alpha^\perp$ at $x = L$ that is implemented by the singular half line, and we are back in the case of a problem on a bounded interval that we already discussed, before formulating Theorem 3.21. That is also true in the other case, when $L \leq 0$, since we can artificially view the right half line problem as a problem on a bounded interval $(0, L')$ by introducing the boundary condition $\alpha^\perp$ at $x = L'$; in other words, we can view $m_+ = -\tan\alpha$ both as the m function of the problem on $(0, \infty)$ and as the m function on $(0, L')$, with boundary condition $\alpha^\perp$ at $x = L'$.

Now take a $g \in \mathcal{S}(0)$, with compact support contained in $[L, \infty)$. Since $P_\alpha T(x;t)$ is constant on $x \geq L$, we then have

$$G(t) = cT^*(L;t)e_\alpha,$$

for some $c \in \mathbb{C}$. This is of the same form as the boundary terms from (3.22), so by the argument we gave there, it follows that G represents the zero element of $L^2(\mathbb{R}, \rho_{L_-})$, for any L_-, or at least we are certain about those L_- that are small enough so that (L_-, L) is not a single singular interval. If there are no such L_-, then $\mathbb{R}$ consists of two singular intervals, which is a trivial case that we can easily discuss directly ($\rho = 0$ then), and I will ignore it here.

In other words,

$$\int_{-\infty}^{\infty} G^*(t)\, d\rho_{L_-}(t) G(t) = 0$$

for small L_-, and now we can again introduce a cut-off function φ_T and then send $L_- \to -\infty$ to conclude that $G = 0$ almost everywhere with respect to ρ also. □

An important general property of these spectral representations is that only the absolutely continuous part can have multiplicity two. I will just state the (well-known) result here, and I refer the reader to the literature for the proof.

Theorem 3.22. *Let $m_\pm$ be two generalized Herglotz functions, and if both $m_\pm(z) = a_\pm \in \mathbb{R}_\infty$ are constant, then we also assume that $a_+ \neq -a_-$, so that $M(z)$ is well defined in all cases. Let ρ be the associated measure. Then the singular part of the operator M_t in $L^2(\mathbb{R}, \rho)$ has multiplicity at most* 1.

For canonical systems, it is of course clear that eigenvalues are simple because at each limit point endpoint, there is at most one solution that is in L^2_H. The interesting part of the statement is that the singular continuous spectrum is also simple (if present). This property is not unexpected, but it takes some work to prove it carefully.

The absolutely continuous spectrum can have multiplicity two. This part of the operator is unitarily equivalent to the orthogonal sum of the absolutely continuous parts of the two half line problems. This follows quickly from abstract scattering theory (since cutting the whole line into two half lines amounts to a rank 2 perturbation of the resolvent). Alternatively, it can be shown by hand by an analysis of the matrix $\operatorname{Im} M(t)$ and its multiplicity.

To make these remarks more concrete, introduce the sets (in fact, it is better to view them as equivalence classes of sets, determined up to sets of Lebesgue measure zero)

$$\Sigma_{ac}^\pm = \{t \in \mathbb{R} : \operatorname{Im} m_\pm(t) > 0\}.$$

Then S_{ac} is unitarily equivalent to multiplication by the variable in $L^2(\Sigma_{ac}^+, dt) \oplus L^2(\Sigma_{ac}^-, dt)$. In particular, the (local) multiplicity of the absolutely continuous part can be read off from this: it equals two on $\Sigma_{ac}^+ \cap \Sigma_{ac}^-$, and it is one on the remaining part of the union of these sets.

3.8 Notes

The remarks made at the end of the previous chapter apply here as well: while this material on the spectral representation of our relations is foundational and, as the phrase goes, well known, it is not often discussed explicitly in the literature. My presentation here again (as in Chapter 2) differs somewhat from the traditional one in that I keep the relations all the way through; one could also reduce the Hilbert space by hand, that is, take it to be what in our notation would be $\overline{D(\mathcal{S})} = L^2_H(a,b) \ominus \mathcal{S}(0)$, and then study the *operators* on this space. A treatment of the spectral representation (of the operator part) along these lines is given in [59].

Section 3.1. The basic Definition 3.1 seems very natural, given what we did before (and will do afterwards), but I have not seen it anywhere in the literature and it could be new.

Section 3.2. The important thing about Theorem 3.4 is that it shows that the spectrum is purely discrete. The additional statement that $\sum 1/(1 + E_n^2) < \infty$ is not very accurate: the complex analytic properties of the transfer matrix that we will establish in the next chapter imply that if $N(R)$ denotes the number of eigenvalues in $(0, R)$ or $(R, 0)$ (whichever interval is non-empty), then $N(R)/R$ and $N(-R)/R$ both approach limits as $R \to \infty$, possibly zero. In particular, $|E_n| \gtrsim n$. I mention these results in passing; we will not prove them here.

Section 3.3. Theorem 3.5 is due to de Branges. The conceptually crystal clear proof given here is from [1].

Corollary 3.6 is also quite interesting when one attempts to apply it to a classical equation, say a Schrödinger or Jacobi equation with a limit circle endpoint. It then suggests that this limit circle endpoint really wants to be regular, except that one perhaps has not chosen the right variables yet to make it so; those variables would amount to the rewriting of the equation as a canonical system.

Section 3.4. The material of this section is very classical indeed, going back (for Sturm–Liouville equations) to Weyl's 1910 paper [66], which predates most of the classical functional analysis. The use of transfer matrices and their action on solution vectors, viewed projectively and thus identified with points on $\mathbb{C}_\infty$, is not just optional gadgetry here. This becomes clear as soon as one sees the uninspired computational arguments that those presentations (especially in older sources) that do not rely on this formalism have to give. In this way, the structure of the theory and what makes it work are completely obscured: Weyl theory has nothing to do with what the underlying equation happens to look like; rather, the source from which everything flows is the existence of transfer matrices with the properties stated in Theorem 1.2.

Section 3.5. As I said above, the material of this section is in principle quite standard, but the presentation I give here, with the relations being front and center, could be new as far as some of the details are concerned. The proof of Theorem 3.17 is more challenging for canonical systems than that of its analogs for more classical equations would be since one does not have much a priori information on the t dependence of the functions $F(t) = (Uf)(t)$. I learned the lovely trick of writing $\int |F|^2\, d\rho = \int \frac{1+t^2}{1+t^2} |F|^2\, d\rho$ that helps us out here from the work of de Branges [17]; it is used in a somewhat different context there.

Near the end of the proof, when we show that U maps onto $L^2(\mathbb{R}, \rho)$, the argument seems (and is) quite routine, but note that the limit point assumption is crucial here: if we had limit circle case at infinity, then there would be a huge supply of measures such that U still maps into $L^2(\mathbb{R}, \rho)$ as a partial isometry, but now this space is much larger than $R(U)$. The Nevanlinna parametrization, which is Theorem 6.2 of Chapter 6, will clarify matters completely here.

Section 3.7. This way of setting up a spectral representation of a whole line problem, using the half line m functions and the matrix valued Herglotz function M, is a very standard tool that is in common use for other equations, but this could be the first detailed discussion of this material for canonical systems. Theorem 3.22 is from [31].

4 Transfer matrices and de Branges spaces

4.1 Review of complex analysis

Our work in this chapter will be somewhat different in character from that of the previous chapters. Complex analysis methods will now take center stage. For this reason, I would like to open it with a report, without proofs mostly, on what you will need to be familiar with to follow the discussion of the subsequent sections.

This material may be roughly divided into two parts, but with considerable overlap between them:

(1) entire functions of exponential type;
(2) function theory on the upper half plane $\mathbb{C}^+$.

These topics are quite classical and there are many excellent introductions. Standard references are [8, 46, 47] for the theory of entire functions and [22, 26] for Hardy spaces and the Nevanlinna class. Reference [36] presents an introduction to both topics; this book does not cover nearly as much ground, but the gentler pace and engaging style make it a joy to read. An especially useful source for specifically our needs here is [60]; I recommend it highly for a quick yet thorough introduction to exactly those topics that matter for us here.

Let us now start with the actual overview. First of all, a piece of notation: if $F(z)$ is entire, then we let $F^{\#}(z) = \overline{F(\overline{z})}$, and this function is also entire. We say that an entire function F is of *exponential type* if

$$|F(z)| \le C(\tau)e^{\tau|z|} \quad (z \in \mathbb{C})$$

for some $\tau > 0$, and the infimum of the $\tau > 0$ that work here is called the *type* of F. A function $F \not\equiv 0$ of exponential type may be represented as a *Hadamard product*, as follows: let $m \ge 0$ be the multiplicity of a possible zero at $z = 0$ and list the remaining zeros (if any) as $z_1, z_2, \ldots$, with repetitions according to multiplicity. Then

$$F(z) = z^m e^{a+bz} \prod_n \left(1 - \frac{z}{z_n}\right) e^{z/z_n};$$

the product converges locally uniformly and unconditionally.

The Hadamard factorization may be viewed as a more convenient version of the general Weierstraß factorization, which becomes available for functions of exponential type.

If F is of exponential type, then its type τ may be computed as

$$\tau = \limsup_{R\to\infty} \frac{\log M(R)}{R}, \quad M(R) \equiv \max_{|z|=R} |F(z)|.$$

So far, this just rephrases the definition. But now we can also introduce the *indicator function* of an entire function of exponential type

$$h(\theta) = \limsup_{R\to\infty} \frac{\log |F(Re^{i\theta})|}{R},$$

https://doi.org/10.1515/9783110563238-004

which contains more specific information about how $|F|$ grows when $|z| \to \infty$. The indicator h is a continuous function of θ, and $\tau = \max_\theta h(\theta)$, $\min h \geq -\tau$.

The *Hardy spaces* $H^p = H^p(\mathbb{C}^+)$ on the upper half plane are defined for $0 < p \leq \infty$, and they consist of holomorphic functions $F(z)$ that satisfy certain growth restrictions as z approaches the boundary $\mathbb{R}_\infty$. More precisely, if $p < \infty$, then H^p is the set of holomorphic functions $F : \mathbb{C}^+ \to \mathbb{C}$ that satisfy

$$\sup_{y>0} \int_{-\infty}^{\infty} |F(t+iy)|^p \, dt < \infty, \tag{4.1}$$

and H^∞ contains all bounded holomorphic functions $F : \mathbb{C}^+ \to \mathbb{C}$.

Except for a quick reference to H^1 in a technical step in a later proof, we will only work with the spaces H^2 and H^∞ here, so I will focus on $p = 2, \infty$ exclusively in the remainder of this section. In a moment, I will also introduce the Nevanlinna class N which may be thought of as the limiting case $p \to 0+$ of H^p. This space will actually be quite important for us.

But let us first review some properties of H^p, $p = 2, \infty$. For F in one of these spaces, the limit $F(x) \equiv \lim_{y\to 0+} F(x+iy)$ exists for almost every $x \in \mathbb{R}$, and $F(z)$ can in principle be recovered from its boundary values $F(x)$. In fact, if $F_1(x) = F_2(x)$ on a positive measure set $x \in A \subseteq \mathbb{R}$, then it will already follow that $F_1 = F_2$.

If $F \in H^2$, then $F(x) \in L^2(\mathbb{R})$, and this reconstruction can be done quite explicitly: H^2 functions satisfy the Cauchy type formula

$$\frac{1}{2\pi i} \int_{-\infty}^{\infty} F(t) \frac{dt}{t-z} = \begin{cases} F(z), & z \in \mathbb{C}^+, \\ 0, & z \in \mathbb{C}^-. \end{cases} \tag{4.2}$$

We can therefore identify H^2 with a subspace of $L^2(\mathbb{R})$, and H^2 is a closed subspace and thus a Hilbert space itself, with norm $\|F(z)\|_{H^2} = \|F(x)\|_{L^2(\mathbb{R})}$.

If we subtract the Cauchy formula for $\overline{z}$ from the original one, then we see that H^2 functions also have a Poisson representation

$$F(z) = \frac{1}{\pi} \int_{-\infty}^{\infty} \frac{y}{(t-x)^2 + y^2} F(t) \, dt, \quad z = x + iy \in \mathbb{C}^+. \tag{4.3}$$

The *Nevanlinna class* $N = N(\mathbb{C}^+)$ can be defined as the set of holomorphic functions $F : \mathbb{C}^+ \to \mathbb{C}$ for which the subharmonic function $\log^+ |F(z)|$ has a harmonic majorant, that is, there is a harmonic function $u : \mathbb{C}^+ \to \mathbb{R}$ such that $\log^+ |F(z)| \leq u(z)$. Here, $\log^+$ denotes the positive part of the logarithm, so $\log^+ x = \log x$ if $\log x \geq 0$ and $\log^+ x = 0$ if $\log x < 0$. An equivalent condition, which emphasizes the analogy to the definition of the Hardy spaces, is given by

$$\sup_{y>0} \int_{-\infty}^{\infty} \frac{\log^+ |F(t+iy)|}{t^2 + y^2 + 1} \, dt < \infty. \tag{4.4}$$

Theorem 4.1. *Let $F : \mathbb{C}^+ \to \mathbb{C}$ be a holomorphic function. Then $F \in N$ if and only if there are $F_1, F_2 \in H^\infty$ such that $F = F_1/F_2$.*

For this reason, functions in N are often called functions of *bounded type*. Theorem 4.1 is quite useful, and one important immediate consequence of it is the fact that functions in N also have boundary values $F(x) = \lim_{y\to 0+} F(x+iy)$ almost everywhere. If $F \not\equiv 0$, then this function will satisfy

$$\int_{-\infty}^{\infty} \frac{|\log|F(t)||}{1+t^2}\, dt < \infty.$$

Another consequence of Theorem 4.1 that will be important for us later is that if F is a Herglotz function, then $F \in N$. Indeed, in this case, we can take

$$F_1(z) = 1 - \frac{i}{F(z)+i}, \quad F_2(z) = \frac{1}{F(z)+i}.$$

Moreover, N contains H^p for any $p > 0$, and, in particular, $N \supseteq H^2, H^\infty$.

If $F \in N$, $F \not\equiv 0$, then its zeros $z_n \in \mathbb{C}^+$ (if any) satisfy the *Blaschke condition*

$$\sum \frac{y_n}{x_n^2 + y_n^2 + 1} < \infty, \quad z_n = x_n + iy_n. \tag{4.5}$$

Given z_ns satisfying (4.5), we can define the corresponding *Blaschke product* as

$$B(z) = \left(\frac{z-i}{z+i}\right)^m \prod_{z_n \neq i} \frac{|z_n^2+1|}{z_n^2+1} \frac{z-z_n}{z-\overline{z_n}}.$$

The precise definition of $B(z)$ is actually not important for us; what matters is that given z_ns satisfying the Blaschke condition, we have a $B \in H^\infty$, $|B(z)| \le 1$, with zeros precisely at the z_ns, and if these were the zeros of an $F \in N$ and we define $F_0(z) = F(z)/B(z)$, then $F_0 \in N$ also.

Now $\log|F_0(z)|$ is a harmonic function on $\mathbb{C}^+$, and the existence of a harmonic majorant of $\log^+|F_0(z)|$ implies (in fact, is equivalent to) a Poisson representation formula for $\log|F_0(z)|$; this step is quite similar to the representation of Herglotz functions. As in that case, we can then easily provide a harmonic conjugate and thus obtain a representation formula for $\log F_0(z)$ itself, and then we exponentiate and obtain a representation of a general $F \in N$, $F \not\equiv 0$, as what is called a *canonical product*

$$F(z) = e^{i\alpha} B(z) \exp\left(-i \int_{\mathbb{R}_\infty} \frac{1+tz}{t-z}\, d\nu(t)\right); \tag{4.6}$$

here, ν is a finite signed measure. The data $e^{i\alpha}, \{z_n\}, \nu$ are uniquely determined by $F \in N$, and it is also true that, conversely, if z_n is any sequence satisfying (4.5) and ν is a finite signed Borel measure on $\mathbb{R}_\infty$, then (4.6) will define a function $F \in N$.

The condition that the positive part ν_+ does not have a singular part defines a subclass N^+ of N. We have $H^2, H^\infty \subseteq N^+$. A partial converse is the statement that if $F \in N^+$ and $F(x) \in L^2(\mathbb{R})$, then $F \in H^2$.

The following result provides an interesting link between functions of exponential type and complex analysis on the upper half plane.

Theorem 4.2 (Krein). *For an entire function F, the following are equivalent:*

(a) *F is of exponential type and*

$$\int_{-\infty}^{\infty} \frac{\log^+ |F(t)|}{1+t^2}\, dt < \infty;$$

(b) $F, F^\# \in N$.

For such functions, we can develop a rather detailed picture of their large $|z|$ asymptotics. We start with a definition.

Definition 4.1. For $F \in N$ with canonical factorization (4.6), we define its *mean type* as $\tau = \tau(F) := \nu(\{\infty\}) \in \mathbb{R}$.

As has been noted by many authors, the term *mean type* for this quantity is not a particularly fitting description, but it seems to be commonly accepted now.

Recall that we also denoted the type of an entire function of exponential type by τ earlier, and this is deliberate: the point mass of ν at infinity contributes a factor $e^{-i\tau z}$ to (4.6), and it is not hard to show that this contribution dominates everything else in the sense that it determines the indicator function $h(\theta)$.

In particular, if we now have an F as in Theorem 4.2, then we can use condition (b) to define two mean types $\tau_+ = \tau(F)$, $\tau_- = \tau(F^\#)$.

Theorem 4.3. *Suppose that $F \not\equiv 0$ satisfies the conditions from Theorem 4.2. Then $h(\theta) = \tau_+ \sin\theta$ for $0 \le \theta \le \pi$ and $h(\theta) = \tau_- |\sin\theta|$ for $\pi \le \theta < 2\pi$.*

The canonical factorization (4.6) simplifies for an entire $F \not\equiv 0$. In general, $\pi(1+t^2)\, d\nu$ may be obtained as the weak-$*$ limit of the measures $\log|F(t+iy)|\, dt$; for an entire function, $F(t+iy)$ is now extremely well behaved when we send $y \to 0+$. This rules out a singular part of ν other than a possible point mass at ∞. We have

$$d\nu(t) = \frac{1}{\pi} \frac{\log|F(t)|}{1+t^2}\, dt + \tau\delta_\infty. \tag{4.7}$$

4.2 De Branges spaces

Definition 4.2. A *de Branges function* is an entire function E satisfying $|E(z)| > |E^\#(z)|$ for $z \in \mathbb{C}^+$. Given a de Branges function E, we define the *de Branges space* $B(E)$ based on E as

$$B(E) = \{F : F \text{ entire}, F/E, F^{\#}/E \in H^2\}.$$

Note that a de Branges function cannot have zeros in $\mathbb{C}^+$, and the ones we will be interested in later do not have zeros on $\mathbb{R}$ either, though that cannot be ruled out for general de Branges functions.

We will often decompose de Branges functions by taking real and imaginary parts on the real line (but keeping all functions entire), so will write $E(z) = A(z) - iC(z)$, with $A = (E + E^{\#})/2$, $C = (i/2)(E - E^{\#})$. (The perhaps slightly unexpected minus sign and the use of the symbol C, rather than B, will be convenient later, when we extract de Branges functions from transfer matrices.)

Theorem 4.4. *Let E be a de Branges function. Then $B(E)$ becomes a Hilbert space when endowed with the scalar product*

$$[F, G] = \frac{1}{\pi} \int_{-\infty}^{\infty} \overline{F(t)} G(t) \frac{dt}{|E(t)|^2}.$$

The reproducing kernels

$$J_w(z) = \frac{\overline{E(w)}E(z) - \overline{E^{\#}(w)}E^{\#}(z)}{2i(\overline{w} - z)} = \frac{\overline{C(w)}A(z) - \overline{A(w)}C(z)}{\overline{w} - z} \tag{4.8}$$

are in $B(E)$, and $[J_w, F] = F(w)$ for all $F \in B(E)$, $w \in \mathbb{C}$.

The slightly unusual square bracket notation for the scalar product will be helpful when we consider de Branges spaces together with other Hilbert spaces (whose scalar products will be denoted by $\langle \cdot, \cdot \rangle$) later.

Proof. We begin by verifying that $J_w \in B(E)$. Clearly, J_w is an entire function of z. To show that $J_w(z)/E(z) \in H^2$, notice that this function is holomorphic on $\mathbb{C}^+$ since $E \neq 0$ there. I now want to make sure that we need not worry about the z close to $\overline{w}$ when discussing (4.1). Now certainly there are no problems if $\overline{w} \in \mathbb{C}^-$, and there also are none if $\overline{w} \in \mathbb{C}^+$ because then J_w/E is bounded on a neighborhood of $\overline{w}$. The only concern really is the case of a real w with $E(w) = 0$, but this case is completely trivial because then $J_w \equiv 0$.

So, when verifying (4.1), we may now restrict the integrations to $|t + iy - \overline{w}| \geq \delta$, for some $\delta > 0$. We then look separately at the two terms coming from (4.8). The first one, as a function of z, is of the form $c/(\overline{w} - z)$, and this certainly satisfies (4.1). The second one is of the same form, but with an extra factor $E^{\#}(z)/E(z)$, but this is bounded by 1 and thus only helping matters.

In the same way, we can show that $J_w^{\#}/E \in H^2$. Thus $J_w \in B(E)$ for all $w \in \mathbb{C}$, as claimed.

We now check that J_w is a reproducing kernel. Let $F \in B(E)$. The Cauchy formula (4.2), applied to F/E and $F^\#/E$, gives

$$\frac{E(w)}{2\pi i}\int_{-\infty}^{\infty}\frac{F(t)}{E(t)}\frac{dt}{t-w} = \begin{cases} F(w), & w \in \mathbb{C}^+, \\ 0, & w \in \mathbb{C}^-, \end{cases}$$

$$\frac{\overline{E^\#(w)}}{2\pi i}\int_{-\infty}^{\infty}\frac{F^\#(t)}{E(t)}\frac{dt}{t-\overline{w}} = \begin{cases} 0, & w \in \mathbb{C}^+, \\ \overline{F(w)}, & w \in \mathbb{C}^-. \end{cases}$$

Add the complex conjugate of the second formula to the first one. This shows that $[J_w, F] = F(w)$ for $w \in \mathbb{C} \setminus \mathbb{R}$. The same statement then follows for all $w \in \mathbb{C}$ because both sides are entire functions of w. To prove that $[J_w, F]$ is a holomorphic function of w, we use a typical argument that is quite routine and perhaps does not need much comment. However, since we will also need it several times later on, let me actually present it in detail here. We want to show that

$$G(w) = \int_{-\infty}^{\infty}\overline{J_w(t)}F(t)\frac{dt}{|E(t)|^2}$$

is holomorphic. By Morera's theorem, we can establish this by showing that $\int_\gamma G(w)\,dw = 0$, for any closed path $\gamma \subseteq \mathbb{C}$. Since $\overline{J_w(t)}$ is a holomorphic function of $w \in \mathbb{C}$ for any fixed t, it has this property of vanishing line integrals, and so we would be done if we could do the w integration first. In other words, we only need to justify changing the order of integration in the double integral $\int_\gamma G(w)\,dw$, and this is very easily accomplished, with the help of Fubini's theorem: w varies over a compact set, and thus we still have the bounds $|J_w(t)/E(t)| \lesssim \min\{1, 1/(1+|t|)\}$ available, and $|F/E| \in L^2$.

It remains to show that $B(E)$ is complete. Since $B(E)$ can be viewed as a subspace of $L^2(\mathbb{R}, dt/(\pi|E(t)|^2))$, this is equivalent to the statement that $B(E)$ is closed in this space. (I should perhaps point out that the measure $dt/|E|^2$ is not guaranteed to be finite on compact sets since E could have real zeros, but that does not cause any problems here.) So let $F_n \in B(E)$, $F_n \to F \in L^2(\mathbb{R}, dt/(\pi|E|^2))$. Since H^2 becomes a closed subspace of $L^2(\mathbb{R})$ after identifying H^2 functions with their boundary values on $\mathbb{R}$, it is clear that $F/E, F^\#/E \in H^2$ (currently F is an element of $L^2(\mathbb{R}, dt/(\pi|E|^2))$, so we interpret $F^\# = \overline{F}$ for now). Moreover, $F \in L^2(\mathbb{R}, dt/(\pi|E|^2))$ is indeed the restriction of an entire function F_0, which we can provide explicitly as $F_0(z) = [J_z, F]$. □

The most basic (non-polynomial) example of a de Branges function is given by $E(z) = e^{-iLz}$, with $L > 0$. To work out what $B(E)$ is, it is useful to recall that H^2 can also be described as the space of Fourier transforms

$$F(z) = \widehat{f}(z) = \int_{-\infty}^{\infty} f(t)e^{itz}\,dt$$

of functions $f \in L^2(0,\infty)$. Thus, for an entire function F, the conditions that F/E, $F^\#/E \in H^2$ are now equivalent to F being the Fourier transform of a function $f \in L^2(-L,L)$. This is the classical Paley–Wiener space PW_L. By the Paley–Wiener theorem, $F \in B(E) = PW_L$ precisely if F is entire of exponential type $\le L$ and its restriction to $\mathbb{R}$ lies in $L^2(\mathbb{R})$. The reproducing kernel of this space is the Dirichlet kernel

$$J_w(z) = \frac{\sin L(\overline{w}-z)}{\overline{w}-z} = D_L(\overline{w}-z),$$

and since (in general) convolution with D_L projects onto the frequencies in $(-L,L)$, we can also see directly that this works as a reproducing kernel.

General de Branges spaces share the essential features of this example. They can always be viewed as spaces of transforms, with the frequency restricted to an interval. In general, however, the transform will use the solutions $u(x,z)$ of a canonical system $Ju' = -zHu$ instead of the exponentials e^{itz}. We will encounter this type of scenario soon, in the next section.

The de Branges function E is not uniquely determined by the space; it may be transformed by an $\mathrm{SL}(2,\mathbb{R})$ matrix, as spelled out in Theorem 4.5. To be able to formulate this concisely, let me introduce the notation $B(E_1) \equiv B(E_2)$ for the statement that the two spaces are isometrically equal to one another. More explicitly, they contain the same functions, and the scalar product of any two functions is the same in both spaces.

Theorem 4.5. *Let E_1, E_2 be de Branges functions. Then $B(E_1) \equiv B(E_2)$ if and only if*

$$\begin{pmatrix} A_2(z) \\ C_2(z) \end{pmatrix} = M \begin{pmatrix} A_1(z) \\ C_1(z) \end{pmatrix} \quad \textit{for some } M \in \mathrm{SL}(2,\mathbb{R}). \tag{4.9}$$

Proof. I first claim that $B(E_1) \equiv B(E_2)$ if and only if the reproducing kernels agree, that is, $J_w^{(1)}(z) = J_w^{(2)}(z)$ for all $w,z \in \mathbb{C}$. This is certainly the case if $B(E_1) \equiv B(E_2)$. Conversely, assume that the reproducing kernels are the same. The linear combinations of the J_w are dense because an element orthogonal to all J_w will be the zero function. Moreover, the scalar products of the J_w among themselves, $[J_w, J_z] = J_z(w)$, do not depend on the space. So the identity map on this dense subspace $L(\{J_w\})$ is an isometry onto a dense subspace again. Its extension to the whole space is still the identity map, and thus $B(E_1) \equiv B(E_2)$.

Suppose now that the reproducing kernels of the two spaces are the same. Then, by (4.8),

$$\overline{C_1(w)}A_1(z) - \overline{A_1(w)}C_1(z) = \overline{C_2(w)}A_2(z) - \overline{A_2(w)}C_2(z) \tag{4.10}$$

for all $w,z \in \mathbb{C}$. The two functions $A_j(z)$, $C_j(z)$ cannot be linearly dependent (as functions) because that would make $|E^\#| = |E|$. So if we view the expressions from (4.10) as functions of z for fixed w and then try various ws, we can produce two linearly independent functions. Thus $A_2(z)$, $C_2(z)$ span the same (two-dimensional) space as $A_1(z)$,

$C_1(z)$, and hence (4.9) indeed follows, at this point for some invertible matrix M. Since A_j, C_j are real on the real line, M must have real entries. It is then easy to verify that (4.10) will also imply that $\det M = 1$, as claimed.

Conversely, if (4.9) holds, then (4.10) is an immediate consequence and thus the reproducing kernels agree. □

4.3 Spectral representation and de Branges spaces

We now come to the main reason for our interest in de Branges spaces: they provide a function theoretic view of the spectral representation of canonical systems with two regular endpoints. This reinterpretation of (some of) the material of Chapter 3 is often useful, especially in inverse spectral theory, when we want to reconstruct the canonical system from its spectral data.

Consider a canonical system on $[0, L]$, and, as always, we assume that this is not a single singular interval.

Theorem 4.6. *The function $E_L(z) = u_1(L, z) - iu_2(L, z)$ is a de Branges function. The reproducing kernel of this space is given by*

$$J_w(z) = \int_0^L u^*(x, w)H(x)u(x, z)\, dx.$$

Proof. We start with the formula for J_w, which follows from a computation:

$$\begin{aligned} J_w(z) &= \frac{\overline{E_L(w)}E_L(z) - \overline{E_L^{\#}(w)}E_L^{\#}(z)}{2i(\overline{w} - z)} = \frac{1}{\overline{w} - z}u^*(L, w)Ju(L, z) \\ &= \int_0^L u^*(x, w)H(x)u(x, z)\, dx; \end{aligned}$$

as usual, the last step is by looking at $(u^*Ju)'$. In particular, taking $w = z$, we see that

$$\frac{|E_L(z)|^2 - |E_L^{\#}(z)|^2}{4 \operatorname{Im} z} = \|u(\cdot, z)\|^2,$$

so E_L is a de Branges function, at least if $\|u(\cdot, z)\| > 0$, but this will hold if $(0, L)$ is not a single singular interval. □

The defining formula for the partial isometry $U : L_H^2(0, L) \to L^2(\mathbb{R}, \rho_L^{(\beta)})$ from Theorem 3.16 that sets up the spectral representation of $\mathcal{S}_L^{(\beta)}$ makes sense for arbitrary $t \in \mathbb{C}$, and we could wonder what collection of functions is then obtained. This has a neat answer, as we will now show: the reinterpreted U maps onto $B(E_L)$.

I will use the notation $\mathcal{S}_0$ for the restriction of $\mathcal{T}$ by the conditions $f_{0,2}(0) = 0$, $f_0(L) = 0$. This is a one-dimensional symmetric extension of the minimal relation $\mathcal{T}_0$:

we have replaced the condition that $f_0(0) = 0$ (which elements $(f,g) \in \mathcal{T}_0$ have to satisfy) by the boundary condition at $x = 0$, and we have not modified near $x = L$. The deficiency index of $\mathcal{S}_0$ is 1, and its self-adjoint realizations are the $\mathcal{S}_L^{(\beta)}$.

Theorem 4.7. *The formula*

$$(Uf)(z) = \int_0^L u^*(x,\overline{z})H(x)f(x)\,dx \tag{4.11}$$

defines a partial isometry $U : L_H^2(0,L) \to B(E_L)$. More precisely, $N(U) = \mathcal{S}_0(0)$, and U maps $\mathcal{S}_0(0)^\perp$ unitarily onto $B(E_L)$.

We did not have the extra complex conjugate in our definition of U in Chapter 3, and it was not needed there because we only considered real z. Here, of course, the complex conjugation is crucial, in order to make Uf a holomorphic function of $z \in \mathbb{C}$.

Proof. We already showed in the proof of Theorem 3.17 that $Ug \equiv 0$ if $(0,g) \in \mathcal{S}_0$, but let me repeat the calculation here, for your convenience: I will denote by f_0 the representative of $0 \in L_H^2(0,L)$ with $Jf_0' = -Hg$, $f_{0,2}(0) = f_0(L) = 0$. Then

$$\begin{aligned}(Ug)(z) &= \int_0^L u^*(x,\overline{z})H(x)g(x)\,dx = -\int_0^L u^*(x,\overline{z})Jf_0'(x)\,dx \\ &= \int_0^L u^{*\prime}(x,\overline{z})Jf_0(x)\,dx = z\int_0^L u^*(x,\overline{z})H(x)f_0(x)\,dx = 0,\end{aligned}$$

since $Hf_0 = 0$ almost everywhere.

Next, take $f(x) = u(x,w)$, with $w \in \mathbb{C}$, and compute

$$\begin{aligned}(Uf)(z) &= \int_0^L u^*(x,\overline{z})H(x)u(x,w)\,dx \\ &= \int_0^L u^*(x,\overline{w})H(x)u(x,z)\,dx = J_{\overline{w}}(z),\end{aligned}$$

by Theorem 4.6. In particular, $Uf \in B(E_L)$ for any such choice of f.

Fix a boundary condition β at $x = L$. The choice of β is almost arbitrary here, except that if $(0,L)$ ends with a singular interval, then I want to avoid the one value that puts us into the exceptional case (i) of Theorem 2.28(b). In other words, and this will be the key point, I fix a β that will make $\mathcal{S}_L^{(\beta)}(0) = \mathcal{S}_0(0)$ (which is almost automatic). For ease of notation, I will denote this self-adjoint relation simply by $\mathcal{S}$ in the remainder of this proof.

We then use (4.11) to define a map

$$U_0 : L(\{u(\cdot,t) : t \in \sigma(\mathcal{S})\}) \to B(E_L).$$

We have just checked that indeed U_0 maps into $B(E_L)$. Moreover, if $s,t \in \sigma(\mathcal{S})$, then

$$\begin{aligned}[U_0u(\cdot,s), U_0u(\cdot,t)] &= [J_s, J_t] = J_t(s) = \langle u(\cdot,t), u(\cdot,s)\rangle \\ &= \langle u(\cdot,s), u(\cdot,t)\rangle,\end{aligned}$$

by Theorem 4.6 again. So U_0 is isometric and thus so will be its unique continuous extension U_1 to $\overline{L(\{u(\cdot,t)\})} = \mathcal{S}_0(0)^{\perp}$. Moreover, U_1 is still given by (4.11), and this we can confirm by an argument that we have already used a number of times on similar occasions: to compute U_1f, we must approximate $f_n \to f$ with $f_n \in D(U_0)$, and then $U_1f = \lim U_0f_n$, as a norm limit in $B(E_L)$. However, then also $(U_0f_n)(z) \to (U_1f)(z)$ pointwise, thanks to the existence of reproducing kernels, but clearly $(U_0f_n)(z) \to \int_0^L u^*(x,\overline{z})H(x)f(x)\,dx$.

Finally, U_1 maps onto $B(E_L)$ because every J_z is in the image, and an $F \in B(E_L)$ that is orthogonal to all reproducing kernels is identically equal to zero. □

As an easy but very important consequence of this result, we now see that a canonical system delivers a chain of nested de Branges spaces $B(E_x)$.

Theorem 4.8. (a) *Let $L_1 \in (0,L_2)$ be a regular point. Then $B(E_{L_1}) \subseteq B(E_{L_2})$, and the inclusion is isometric.*

(b) *Let $L \in (0,\infty)$ be a regular point, assume limit point case at infinity, and let ρ be the spectral measure of the half line problem that is obtained from its m function. Then $B(E_L) \subseteq L^2(\mathbb{R},\rho)$, and the inclusion is isometric.*

(c) *These statements also hold if $(0,L_1)$ or $(0,L)$ is a singular interval as long as its type is distinct from e_2.*

The situation from part (c) would normally be excluded by our basic assumption that we never consider canonical systems consisting of a single singular interval only, and we would not be interested in it either. We will need this statement in Chapters 5 and 6.

If $(0,L)$ is a singular interval of type e_2, then of course $E_L(z) = 1$ is not even a de Branges function (though, if we absolutely wanted to, we could still salvage the statement by declaring the corresponding de Branges space to be the zero space).

Proof. (a) Let $\mathcal{H}(L) = L^2_H(0,L)\ominus\mathcal{S}_0(0)$ (and $\mathcal{S}_0$ of course also denotes the corresponding relation on $(0,L)$, even though the dependence on L has not been made explicit). By Theorem 4.7, U maps $\mathcal{H}(L)$ unitarily onto $B(E_L)$. Now the desired conclusion is immediate from the observation that $\mathcal{H}(L_1)$ is isometrically contained in $\mathcal{H}(L_2)$. Or, to spell this out more explicitly and also to understand where the assumption that L_1 is regular comes in, note that $\mathcal{H}(L)$ is essentially the space described in Theorem 3.19. More precisely, it is exactly that space if, when choosing the self-adjoint $\mathcal{S}$ in that theorem, we avoid the one exceptional boundary condition β at $x = L$ in the case when $(0, L)$ ends with a singular interval.

This makes obvious my claim that $\mathcal{H}(L_1) \subseteq \mathcal{H}(L_2)$ isometrically, and it also shows that this will not work if L_1 lies in a singular interval (c, d) of type α, say, because then $\mathcal{H}(L_1)$ will contain $\chi_{(c,L_1)}e_\alpha$, which will not be in $\mathcal{H}(L_2)$.

(b) This is proved by the exact same argument, applied to the isometric inclusion $\mathcal{H}(L) \subseteq L^2_H(0,\infty) \ominus \mathcal{S}(0)$, with $\mathcal{S}$ denoting the self-adjoint realization.

(c) This is the same argument, in a trivial version: the (one-dimensional) space $L(\chi_{(0,L)}e_\alpha)$ still sits isometrically inside $\mathcal{H}(L_2)$ or $L^2_H(0,\infty) \ominus \mathcal{S}(0)$, and it is straightforward (though tedious) to check that it is mapped unitarily onto $B(E_L) = L(1)$. □

Theorem 4.9. (a) *If* $(0, L)$ *does not end with a singular interval of type* $\beta^\perp$, *then* $B(E_L) \equiv L^2(\mathbb{R}, \rho_L^{(\beta)})$ *in the sense that the restriction map* $F \mapsto F|_{\mathbb{R}}$ *is unitary between these spaces.*

(b) *If* $(0, L)$ *ends with a singular interval* (c, L) *of type* $\beta^\perp$, *then* $R_\beta \in B(E_L) \setminus \{0\}$, *with*

$$R_\beta(z) = u^*(L, \overline{z})e_{\beta^\perp} = -u_1(L,z)\sin\beta + u_2(L,z)\cos\beta.$$

The restriction operator maps $B(E_L) \ominus L(R_\beta)$ *unitarily onto* $L^2(\mathbb{R}, \rho_L^{(\beta)})$, *and* $R_\beta \mapsto 0 \in L^2(\mathbb{R}, \rho_L^{(\beta)})$.

Proof. (a) The assumption makes sure that the $\mathcal{S}_0$ from Theorem 4.7 satisfies $\mathcal{S}_0(0) = \mathcal{S}(0)$, and here I have again abbreviated $\mathcal{S} = \mathcal{S}_L^{(\beta)}$. Thus the adjoint U^* of the map from that theorem together with the U from Theorem 3.16 gives us a sequence of unitary maps

$$B(E_L) \to \overline{D(\mathcal{S})} \to L^2(\mathbb{R}, \rho_L^{(\beta)}).$$

If we now recall what the two Us actually do on functions $f \in L^2_H(0, L)$, then it becomes immediately clear that the composed map acts by restriction.

This argument, slightly modified, also works in case (b): now the space in the middle becomes

$$\mathcal{S}_0(0)^\perp = \overline{D(\mathcal{S})} \oplus L(\chi_{(c,L)}e_{\beta^\perp}),$$

and the second map annihilates the second summand. It does map the first summand unitarily onto $L^2(\mathbb{R}, \rho_L^{(\beta)})$, and the first map is still unitary. This gives the asserted properties of the restriction map in this case, and its kernel is spanned by

$$(U\chi_{(c,L)}e_{\beta^\perp})(z) = \left(\int_c^L u^*(x,\overline{z})h(x)\,dx\right)e_{\beta^\perp},$$

if $H = hP_{\beta^\perp}$ on (c, L). Since $u^*e_{\beta^\perp}$ is constant on (c, L), this function is a (non-zero) multiple of $u^*(L, \overline{z})e_{\beta^\perp}$, as claimed. Finally, if we had $R_\beta \equiv 0$, then $(u_1/u_2)(L, z)$ would be identically equal to a real constant, which is only possible if $(0, L)$ were a single singular interval. □

The converse of this result is also true: these spectral measures are the only measures that realize the de Branges space $B(E_L)$ as an $L^2(\mathbb{R}, \rho)$ space.

Theorem 4.10. *If ρ is a Borel measure on $\mathbb{R}$ such that $B(E_L) \equiv L^2(\mathbb{R}, \rho)$, then $\rho = \rho_L^{(\beta)}$ for some β.*

Proof. Clearly, such a ρ must be a discrete measure $\rho = \sum g_n \delta_{t_n}$, or else $L^2(\mathbb{R}, \rho)$ would contain functions that do not even have an entire extension, let alone one in $B(E_L)$. It is then also clear that the restrictions of the reproducing kernels must satisfy $J_{t_n}(t_k) = (1/g_n)\delta_{kn}$, and, in particular, $J_{t_n}(t_k) = 0$ for all $k \neq n$. By writing this condition out, we see that

$$e^{i\beta} E_L(t_k) - e^{-i\beta} \overline{E_L(t_k)} = 0 \tag{4.12}$$

for all t_k from the support of ρ, for some β. Since $E_L(z) = u_1(L, z) - iu_2(L, z)$, another quick calculation will now show that (4.12) is equivalent to $e_\beta^* J u(L, t_k) = 0$, which is the condition for t_k to lie in $\sigma(\mathcal{S}_L^{(\beta)})$. Now the support of ρ cannot be a proper subset of $\sigma(\mathcal{S}_L^{(\beta)})$ because the formula from Theorem 4.6 shows that also $J_s(t) = 0$ for $s, t \in \sigma(\mathcal{S}_L^{(\beta)})$, $s \neq t$, so such a J_s would have zero norm in $L^2(\mathbb{R}, \rho)$ if ρ did not give weight to $\{s\}$.

So the support of ρ is exactly the spectrum of $\mathcal{S}_L^{(\beta)}$, for some β, and then the weights must also agree with those of the spectral measure $\rho_L^{(\beta)}$, or again the norms of the reproducing kernels would not come out right. □

Theorem 4.9 also gives us the following supplement to Theorem 4.8(b).

Theorem 4.11. *Assume limit point case at infinity and let ρ be the spectral measure that is obtained from the m function. Then*

$$L^2(\mathbb{R}, \rho) = \overline{\bigcup B(E_L)},$$

with the union taken over all regular points $L > 0$.

Proof. If there are arbitrarily large regular points L, then this is an immediate consequence of Theorem 4.8(b) and its proof, since the union of the spaces $\mathcal{H}(L)$ considered there is dense in $L_H^2(0, \infty) \ominus \mathcal{S}(0)$. If $(0, \infty)$ ends with a singular interval (L, ∞) of type $\beta^\perp$, say, then we do not even need the closure: now $\rho = \rho_L^{(\beta)}$ by Theorem 3.18, L is regular, and $L^2(\mathbb{R}, \rho) \equiv B(E_L)$ by Theorem 4.9, since we certainly cannot be in case (b) now, by the definition of L. □

So we can now think of the spaces $L^2(\mathbb{R}, \rho_L^{(\beta)})$ of the spectral representation of $\mathcal{S}_L^{(\beta)}$ as de Branges spaces $B(E_L)$, and this raises the question of what happens to the multiplication operators M_t if we realize them in these new spaces $B(E_L)$. Before I can formulate the answers, a few preparations will be needed. First of all, on any de Branges space $B(E)$ multiplication by the variable z naturally defines an operator M_z, given by

$$M_z = \{(F,G) : F,G \in B(E), zF(z) = G(z)\}.$$

I have used the language of relations here, but obviously $M_z(0) = 0$, and M_z is an operator.

Next, recall our earlier definition of the relation

$$\mathcal{S}_0 = \{(f,g) \in \mathcal{T} : f_{0,2}(0) = f_0(L) = 0\},$$

which is best thought of as a minimal relation of sorts, but only with respect to the right endpoint L. It is closed and symmetric and has equal deficiency indices $\gamma_+(\mathcal{S}_0) = \gamma_-(\mathcal{S}_0) = 1$, and the self-adjoint extensions of $\mathcal{S}_0$ are exactly the relations $\mathcal{S}_L^{(\beta)}$.

The decomposition from Theorem 2.11 also works for the symmetric relation $\mathcal{S}_0$. Namely, set $\mathcal{H}_1 = \mathcal{S}_0(0)^\perp$, $S_0 = \mathcal{S}_0 \cap (\mathcal{H}_1 \oplus \mathcal{H}_1)$. Then

$$\mathcal{S}_0 = S_0 \oplus (0, \mathcal{S}_0(0)),$$

and S_0 is an *operator* in $\mathcal{H}_1$. The proof is straightforward and only repeats the arguments from the proof of Theorem 2.11. The multi-valued part $(0, \mathcal{S}_0(0))$ is already self-adjoint on $\mathcal{H}_1^\perp = \mathcal{S}_0(0)$, so $\mathcal{S}_0^* = S_0^* \oplus (0, \mathcal{S}_0(0))$. In particular, S_0 is closed and symmetric and also has deficiency index 1. Its self-adjoint extensions give us exactly the self-adjoint extensions of $\mathcal{S}_0$.

The only difference to the self-adjoint case that was treated in Theorem 2.11 is that this time, we cannot be sure if S_0 is densely defined. More precisely,

$$\overline{D(S_0)} = \overline{D(\mathcal{S}_0)} = \mathcal{S}_0^*(0)^\perp \subseteq \mathcal{S}_0(0)^\perp,$$

by Lemma 2.12. We know that $\mathcal{S}_0^*(0) = \mathcal{S}_0(0)$ if and only if $(0, L)$ does not end with a singular interval; we discussed this earlier, but if you have doubts, show it again with the help of Theorem 2.28. If $(0, L)$ does end with a singular interval (c, L) of type β, then $\mathcal{S}_0^*(0) = \mathcal{S}_0(0) \oplus L(\chi_{(c,L)} e_\beta)$, so in this case,

$$\mathcal{H}_1 = \overline{D(S_0)} \oplus L(\chi_{(c,L)} e_\beta).$$

Theorem 4.12. *The relation M_z is closed and symmetric, $M_z(0) = 0$, and it has equal deficiency indices $\gamma_\pm(M_z) = 1$. If $U : L_H^2(0, L) \to B(E_L)$ denotes the map from Theorem 4.7 and U_1 denotes its unitary restriction to $\mathcal{H}_1 = \mathcal{S}_0(0)^\perp$, then*

$$U\mathcal{S}_0 U^* = US_0 U^* = U_1 S_0 U_1^* = M_z. \tag{4.13}$$

It is tempting to express the first three claims by saying that M_z is a closed symmetric operator, but this creates potential for confusion because one normally means by this a closed operator that has to be densely defined to have a unique operator adjoint, which is then assumed to be an extension of the original operator (symmetry). Here, however, I mean a relation that is closed and symmetric, as a relation, and then also an operator, and such a relation need not be densely defined.

Proof. We start with (4.13). The first two equalities are obvious because $\mathcal{S}_0(0)$ is annihilated by U, so the multi-valued part of $\mathcal{S}_0$ does not contribute to the relation product.

We already saw in the proof of Theorem 4.7 that if $(f, g) \in \mathcal{S}_0$, then $Uf \in D(M_z)$ and $M_z Uf = Ug$. Conversely, suppose now that $F \in D(M_z)$ and let $g = U^* M_z F \in \mathcal{S}_0(0)^\perp$. Then define

$$f(x) = -J \int_x^L H(t)g(t)\, dt,$$

so $Jf' = -Hg$, $f(L) = 0$. We compute

$$\begin{aligned} zF(z) &= \int_0^L u^*(x, \overline{z}) H(x) g(x)\, dx = -\int_0^L u^*(x, \overline{z}) Jf'(x)\, dx \\ &= e_1^* Jf(0) + \int_0^L u^{*\prime}(x, \overline{z}) Jf(x)\, dx \\ &= e_1^* Jf(0) + z \int_0^L u^*(x, \overline{z}) H(x) f(x)\, dx. \end{aligned}$$

Taking $z = 0$, we deduce that $e_1^* Jf(0) = 0$, that is, f satisfies the boundary condition at $x = 0$, so $(f, g) \in \mathcal{S}_0$ and $F = Uf$.

We have shown that $UD(\mathcal{S}_0) = D(M_z)$, and, as we observed earlier, if $f \in D(\mathcal{S}_0)$, then $U_1 \mathcal{S}_0 f = M_z U_1 f$. So $U_1 \mathcal{S}_0 U_1^* = M_z$, as claimed.

The other properties of M_z are now immediate from our preliminary discussion of S_0, because S_0 has these and M_z is unitarily equivalent to S_0. □

Theorem 4.13. *The following conditions are equivalent:*
(a) $\overline{D(M_z)} \neq B(E_L)$;
(b) $(0, L)$ *ends with a singular interval* (c, L) *of type* $\beta^\perp$ *for some* β;
(c) $R_\beta = -u_1(L, z) \sin\beta + u_2(L, z) \cos\beta \in B(E_L)$ *for some* β.

If these hold, then the β*s from parts* (b) *and* (c) *agree, and*

$$B(E_L) = \overline{D(M_z)} \oplus L(R_\beta).$$

Note also that it can never be the case that $R_\beta \in B(E_L)$ for two distinct values of β. In that case, we could take linear combinations to conclude that $E_L \in B(E_L)$ also, but this is absurd.

Proof. The equivalence of (a) and (b) is obvious from Theorem 4.12 and the discussion preceding it. Moreover, we also observed there that if (b) holds, then

$$\mathcal{S}_0(0)^\perp = \overline{D(\mathcal{S}_0)} \oplus L(\chi_{(c,L)} e_{\beta^\perp}),$$

and so in particular, $U\chi_{(c,L)}e_{\beta^\perp} \in B(E_L) \ominus \overline{D(M_z)}$. We saw in the proof of Theorem 4.9 that this function is a multiple of R_β, so (c) follows and the β is the one from part (b). We have also established the additional claim of Theorem 4.13; recall here that if $\mathcal{S}_0(0)^\perp \neq \mathcal{S}_0^*(0)^\perp$, then the difference is one-dimensional, and thus the same is true for $B(E_L)$ and $\overline{D(M_z)}$ because these spaces are obtained as images under the unitary map U_1.

We now finish the proof by showing that (c) implies (a). To do this, pick an $s \in \sigma(\mathcal{S}_L^{(\beta)})$. Then $u_1(L,s)/u_2(L,s) = \cot\beta$. Since, in general,

$$J_s(z) = \frac{\overline{C(s)}A(z) - \overline{A(s)}C(z)}{\overline{s} - z},$$

our choice of s makes $(s-z)J_s(z)$ a multiple of $R_\beta(z)$. Thus, if $F \in D(M_z)$, then

$$[R_\beta, F] = k[J_s, (s-z)F] = k(s-z)F(z)|_{z=s} = 0.$$

This says that $R_\beta \in D(M_z)^\perp$. □

Finally, we can now ask ourselves how the self-adjoint realizations $\mathcal{S}_L^{(\beta)}$ fit into this. We know, from Theorem 4.12, that these correspond to the self-adjoint extensions of M_z in $B(E_L)$, and we have various characterizations of those. For example, since $Uu(\cdot,z) = J_{\overline{z}}$, the von Neumann formula for M_z^* becomes

$$M_z^* = M_z \oplus L(J_{-i}, iJ_{-i}) \oplus L(J_i, -iJ_i),$$

(and here formally completely correct notation would be $L((J_{-i}, iJ_{-i}))$, etc., but this seems too pedantic). Now one description of the self-adjoint realizations is provided by Theorem 2.19.

Or, for a more direct approach, recall that if we take any $s \in \sigma(\mathcal{S}_L^{(\beta)})$, then we can obtain this realization as

$$\mathcal{S}_L^{(\beta)} = \mathcal{S}_0 \dotplus L(u(\cdot,s), su(\cdot,s)).$$

Move this over to $B(E_L)$ by applying U_1 to obtain the corresponding self-adjoint extension $M_z^{(\beta)}$ of M_z, given by

$$M_z^{(\beta)} = M_z \dotplus L(J_s, sJ_s) = M_z \dotplus L\left(\frac{R_\beta(z)}{s-z}, \frac{sR_\beta(z)}{s-z}\right).$$

In the second equality, I have again made use of the identity $J_s = kR_\beta/(s-z)$, which is valid for $s \in \sigma(\mathcal{S}_L^{(\beta)})$ and was observed in the proof of Theorem 4.13.

If $(0,L)$ ends with a singular interval (c,L) of type $\beta^\perp$, then, as we have experienced many times now, the boundary condition β at $x = L$ is special. We can confirm this one more time here: in this case, $u(\cdot, s) \in D(\mathcal{S}_0)$ and $J_s \in D(M_z)$ already. Let me check the second version of this statement. We have

$$z\frac{R_\beta(z)}{s-z} = -R_\beta(z) + s\frac{R_\beta(z)}{s-z} \in B(E_L),$$

as claimed. So we are actually not extending the domain; the extension of the relation happens by extending the multi-valued part. More specifically, since both (J_s, sJ_s) and (J_s, zJ_s) are in $M_z^{(\beta)}$, we have

$$M_z^{(\beta)}(0) = L((s-z)J_s(z)) = L(R_\beta).$$

4.4 Transfer matrices

Definition 4.3. We define TM as the collection of matrix functions $T : \mathbb{C} \to \mathrm{SL}(2,\mathbb{C})$ with the following properties: $T(z)$ is entire, $T(0) = 1$, $T^\# = T$, and if $\operatorname{Im} z \geq 0$, then

$$i(T^*(z)JT(z) - J) \geq 0. \tag{4.14}$$

We will denote the matrix elements of such a T by

$$T(z) = \begin{pmatrix} A(z) & B(z) \\ C(z) & D(z) \end{pmatrix}.$$

The condition that $T^\# = T$ refers to these entries, so we ask that $F^\# = F$ for $F = A, B, C, D$. This is equivalent to $T(x)$ having real entries for $x \in \mathbb{R}$.

Definition 4.3 lists the general properties that we established for transfer matrices $T(z) = T(L; z)$ of a canonical system on $[0, L]$ in Theorem 1.2, so such a $T(z) = T(L; z)$ lies in TM. In this section, we will study the function theoretic properties of matrix functions $T(z) \in TM$. As I already pointed out in Chapter 1, this is not really more general than just studying transfer matrices $T(z) = T(L; z)$, but it is not a pointless deliberate complication either because we will need the abstract results of this section to eventually prove this major result that every $T \in TM$ is the transfer matrix of some canonical system.

We begin with a discussion of condition (4.14). Please also recall Lemma 3.9, which said that this condition is equivalent to $w \mapsto T^{-1}(z)w$ being a Herglotz function for $\operatorname{Im} z \geq 0$.

Lemma 4.14. *If* $M = \left(\begin{smallmatrix} a & b \\ c & d \end{smallmatrix}\right) \in \mathrm{SL}(2,\mathbb{C})$ *satisfies* $i(M^*JM - J) \geq 0$, *then so do the following matrices:*

$$M_1 = \begin{pmatrix} d & b \\ c & a \end{pmatrix}, \quad M_2 = \begin{pmatrix} a & -c \\ -b & d \end{pmatrix}, \quad M_3 = \begin{pmatrix} \overline{d} & -\overline{b} \\ -\overline{c} & \overline{a} \end{pmatrix}, \quad M_4 = M^*.$$

Of course, these statements can be combined, and the same property for example also follows for $M_5 = \left(\begin{smallmatrix} d & -c \\ -b & a \end{smallmatrix}\right)$.

Proof. (1) If $w \in \mathbb{C}^+$, then $-w \in \mathbb{C}^-$, so $M(-w) \in \mathbb{C}^-$ as well, and thus $-M(-w) \in \mathbb{C}^+$. Equivalently (by Lemma 3.9), the matrix $M_1 = IM^{-1}I$, with $I = \left(\begin{smallmatrix} 1 & 0 \\ 0 & -1 \end{smallmatrix}\right)$, satisfies (4.14).

(2) Multiply (4.14) for M by $J^* = -J$ and J from the left and right, respectively. This shows that

$$i((-JMJ)^* J(-JMJ) - J) \geq 0,$$

so the condition holds for $M_5 = -JMJ$, and then also for M_2 by combining this with part (1).

(3) Take complex conjugates in (4.14) to see that $-i(M^t J\overline{M} - J) \geq 0$. Then pull out $M^t = \overline{M}^*$ and $\overline{M}$; this gives $i(\overline{M}^{-1*} J\overline{M}^{-1} - J) \geq 0$. So $M_3 = \overline{M}^{-1}$ has the asserted property.

(4) Finally, the claim on M_4 follows by combining the first three statements. □

Next we observe that the two entries of a row or column of a $T(z) \in TM$ can never be zero simultaneously at any $z \in \mathbb{C}$ since $\det T(z) = 1$. Thus the quotients A/B, etc. are well defined if they are interpreted as points on the Riemann sphere $\mathbb{C}_\infty$.

Lemma 4.15. *For $z \in \mathbb{C}^+$, we have*

$$-\frac{A(z)}{C(z)}, -\frac{B(z)}{D(z)}, \frac{A(z)}{B(z)}, \frac{C(z)}{D(z)} \in \overline{\mathbb{C}^+}$$

and $|E(z)| \geq |E^\#(z)|$, with $E(z) = A(z) - iC(z)$.

Proof. This is an immediate consequence of (4.14), combined with Lemma 3.9. We know that the linear fractional transformation $w \mapsto T^{-1}(z)w$ is a Herglotz function for $z \in \mathbb{C}^+$, and

$$T^{-1} = \begin{pmatrix} D & -B \\ -C & A \end{pmatrix}.$$

Taking $w = 0, \infty$ shows that $-B/A, -D/C \in \overline{\mathbb{C}^+}$, and thus also $A/B, C/D \in \overline{\mathbb{C}^+}$, as claimed. Now use Lemma 4.14 to switch D, A in this matrix, and then we see in the same way that $-A/C, -B/D \in \overline{\mathbb{C}^+}$.

The inequality on E is equivalent to $\operatorname{Im}(-A/C) \geq 0$. □

Lemma 4.16. *$A(z), D(z) \neq 0$ for $z \in \mathbb{C} \setminus \mathbb{R}$, and if B has a non-real zero, then $B \equiv 0$. The same statement holds for C.*

Proof. Since all four functions $F = A, B, C, D$ are real on the real line, non-real zeros come in complex conjugate pairs. So if F had a non-real zero, then also $F(z_0) = 0$ for some $z_0 \in \mathbb{C}^+$. This means that the generalized Herglotz function $\pm F/G$ that we form with an appropriate partner $G = A, B, C, D$, as in Lemma 4.15, takes the value 0 on $\mathbb{C}^+$, so $F \equiv 0$.

If $F = A$ or $F = D$, then this is impossible since $F(0) = 1$. □

Motivated by these observations, we now introduce the slightly smaller class

$$TM_1 = \{T \in TM : B, C \not\equiv 0\}.$$

Theorem 4.17. *Let $T \in TM$. Then $T \notin TM_1$ if and only if*

$$T(z) = \begin{pmatrix} 1 & -az \\ 0 & 1 \end{pmatrix} \quad \text{or} \quad T(z) = \begin{pmatrix} 1 & 0 \\ az & 1 \end{pmatrix}$$

for some $a \ge 0$.

Proof. It is very easy to verify that these matrices are in TM, and obviously they are not in TM_1.

Conversely, suppose that $T \in TM$, $C \equiv 0$. So $T = \left(\begin{smallmatrix} A & B \\ 0 & 1/A \end{smallmatrix}\right)$, and $T^{-1}(z)w = w/A^2(z) - B(z)/A(z)$. This must be a Herglotz function of w for every $z \in \mathbb{C}^+$. This implies that $A^2(z) > 0$ for $z \in \mathbb{C}^+$, and thus $A \equiv 1$. It then follows that $-B(z)$ has to be a generalized Herglotz function, but B is also entire and real on the real line. It is an immediate consequence of the Herglotz representation theorem, together with the description of the associated measure, that the only such functions are $-B(z) = b + az$, $a \ge 0$. Since $B(0) = 0$, we have $b = 0$ here.

The other case, when $B \equiv 0$, is similar. □

Let us summarize what we have so far, and I will formulate this for $T \in TM_1$, which will give us slightly more elegant statements.

Theorem 4.18. *Let $T \in TM_1$. Then A, B, C, D have only real zeros. The quotients $-A/C$, $-B/D$, A/B, C/D are Herglotz functions, and $E = A - iC$ is a de Branges function.*

Proof. We showed most of this already, including the claim on the zeros. It only remains to prove that the inequalities implicit in Lemma 4.15 become strict, so for example $\operatorname{Im}(-A(z)/C(z)) > 0$ now for $z \in \mathbb{C}^+$. Equality at a single $z_0 \in \mathbb{C}^+$ would imply that $A(z) = aC(z)$ for some $a \in \mathbb{R}$ for all $z \in \mathbb{C}$, but we only need to take $z = 0$ now to see that this is impossible. The other statements have similar proofs. □

Given our work in this chapter and the previous one, it is now natural to wonder if these functions have any spectral theoretic meaning if $T(z) = T(L; z)$ is a transfer matrix of a canonical system. That is definitely the case for $E = u_1 - iu_2$, and we studied this de Branges function and the associated space in detail in the previous section. The quotients $-B/A = -v_1(L,z)/u_1(L,z)$ and $-D/C = -v_2/u_2$ are the familiar m functions $m_L^{(\beta)}(z)$ for $\beta = \pi/2$ and $\beta = 0$, respectively. Finally, $-A/C = -u_1/u_2$ and $-B/D = -v_1/v_2$ may be viewed as mirror versions of these, with the roles of the endpoints switched: we now start out with a solution that satisfies a given boundary condition at $x = 0$ and that could be u or v (or something else), depending on what this boundary condition is actually equal to, and then we evaluate the quotient at $x = L$.

Let us return to the abstract general investigation of the matrix functions $T \in TM$. With the above preparations in place, we can now prove a first more substantial result.

Theorem 4.19. *Let $T \in TM_1$. Then $A, B, C, D, E \in N$, and all five functions share the same mean type $\tau \geq 0$. Moreover, $1/E \in N$, and in fact we have the stronger statement that*

$$\frac{1}{(z+i)E(z)} \in H^2.$$

It is convenient to state the theorem in this compact fashion, but the reader should also be aware that quite a few more specific statements about the properties of these functions are immediate consequences, and let me perhaps elaborate on this before proving it.

First of all, recall Definition 4.1 in this context, for the meaning of $\tau = \tau(F)$ of a function $F \in N$. Since $F = F^{\#}$ for $F = A, B, C, D$, Theorem 4.19 shows that these four functions satisfy condition (b) of Theorem 4.2. Hence they are of exponential type, and so is $E = A - iC$. Moreover, Theorem 4.3 applies: the indicator function of $F = A, B, C, D$ equals $h(\theta) = \tau|\sin\theta|$. Since there are no zeros off the real line, we can be more specific still:

$$\lim_{R\to\infty} \frac{\log|F(Re^{i\theta})|}{R} = \tau|\sin\theta|, \quad F = A, B, C, D, \ \theta \neq 0, \pi. \tag{4.15}$$

For real arguments, we know that

$$\limsup_{R\to\infty} \frac{\log|F(\pm R)|}{R} = 0, \qquad \int_{-\infty}^{\infty} \frac{|\log|F(t)||}{1+t^2}\, dt < \infty.$$

For all five functions, τ is not just the mean type of the restriction to $\mathbb{C}^+$, but also the type of the entire function.

As for E, we also have property (4.15), but only in the upper half plane $0 < \theta < \pi$; in the lower half plane, one needs $\tau(E^{\#})$ instead of τ and, due to the possible presence of zeros, the limit needs to be changed back to a lim sup. We also know that

$$\int_{-\infty}^{\infty} \frac{1}{t^2+1} \frac{dt}{|E(t)|^2} < \infty, \qquad \int_{-\infty}^{\infty} \frac{\log^+|E(t)|}{t^2+1}\, dt < \infty,$$

so $E(t)$ can neither get very small nor very large on average. Finally, see also Theorem 4.23 below for another interesting consequence of Theorem 4.19.

Proof. By (4.14),

$$M(z) \equiv T^{-1}(z)i = -\frac{B(z) - iD(z)}{A(z) - iC(z)}$$

is a Herglotz function. Now recall that if $v = (w, 1)^t$, then $-iv^*Jv = 2\,\mathrm{Im}\, w$. Also, observe that

$$T^{-1}(z)\begin{pmatrix} i \\ 1 \end{pmatrix} = E(z)\begin{pmatrix} M(z) \\ 1 \end{pmatrix}.$$

So, if we now take $v = (i, 1)^t$, then, for $z \in \mathbb{C}^+$, we have

$$2|E(z)|^2 \operatorname{Im} M(z) = -iv^* T^{-1*}(z) J T^{-1}(z) v \geq -iv^* J v = 2.$$

The Herglotz function M is meromorphic, with all poles in $\mathbb{C}^-$, so the associated measure is purely absolutely continuous, and

$$\operatorname{Im} M(z) = by + \frac{1}{\pi} \int_{-\infty}^{\infty} \frac{y}{(t-x)^2 + y^2} \operatorname{Im} M(t)\, dt, \quad z = x + iy, \tag{4.16}$$

and here we also know that

$$\int_{-\infty}^{\infty} \frac{\operatorname{Im} M(t)}{t^2 + 1}\, dt < \infty. \tag{4.17}$$

It might also be helpful to point out that $\operatorname{Im} M(t) = 1/|E(t)|^2$ for $t \in \mathbb{R}$, though we are not going to use this here. It is now routine to verify, using (4.16), (4.17), that

$$\sup_{y>0} \int_{-\infty}^{\infty} \frac{\operatorname{Im} M(t+iy)}{t^2 + (y+1)^2}\, dt < \infty,$$

and since $1/|E|^2 \leq \operatorname{Im} M$, this establishes that $1/[(z+i)E(z)] \in H^2$, as claimed.

Products and quotients of functions from N are in N also (if they are holomorphic on $\mathbb{C}^+$), by Theorem 4.1, and obviously $z + i \in N$. This confirms that $1/E, E \in N$, as claimed. The mean type of $1/E$ has to be non-positive, or we would obtain a contradiction to the H^2 property just established. Or, in more concrete style, you could deduce this from (4.16) and the inequality $1/|E|^2 \leq \operatorname{Im} M$. So $\tau = \tau(E) = -\tau(1/E) \geq 0$.

It remains to prove that $F \in N$ and $\tau(F) = \tau$ also for $F = A, B, C, D$. This follows quickly from the Herglotz property of the quotients. We can write

$$E = A - iC = A\left(1 - i\frac{C}{A}\right) = -iC\left(1 + i\frac{A}{C}\right),$$

and since $\operatorname{Im}(C/A) > 0$ on $\mathbb{C}^+$, this shows that $|A|, |C| < |E|$ there, so $A, C \in N$ as well; for this step, use the original definition of N, in terms of harmonic majorants. Then also $B = A(B/A) \in N$, $D = C(D/C) \in N$. Finally, Herglotz functions have subexponential asymptotics, and thus such a factor cannot affect the mean type, so all five functions share the same $\tau(F)$. To spell this out more explicitly, since these functions are zero-free on $\mathbb{C}^+$, you could compute the mean type as

$$\tau(F) = \lim_{y \to \infty} \frac{\log |F(iy)|}{y},$$

and if $Q(z)$ is a Herglotz function, then $|Q(iy)|, 1/|Q(iy)| = O(y)$ as $y \to \infty$. □

We now show that the H^2 condition on E actually characterizes the de Branges functions that come from a matrix $T \in TM$.

Theorem 4.20. *Let E be a de Branges function and suppose that $1/[(z+i)E(z)] \in H^2$, $E(0) = 1$. Write $E = A - iC$, as usual, with A, C entire and real on the real line. Then there is a matrix function $T \in TM$ whose first column is $(A, C)^t$.*

Proof. We need to provide two more functions B, D that combine with A, C to form a $T \in TM$. To do this, we use the same device that got the previous proof started. Namely, if we had B, D already, then for any $q \in \mathbb{C}^+$, the function

$$T^{-1}(z)q = \begin{pmatrix} D & -B \\ -C & A \end{pmatrix} q = -\frac{B - qD}{A - qC} \tag{4.18}$$

would be a Herglotz function. Now the point is that we can reconstruct these functions without knowing what B, D are (and then we will reverse engineer B, D from that). This is so because $T^{-1}(z)q$ is meromorphic with all poles in $\mathbb{C}^-$, so the measure from its Herglotz representation is $(1/\pi)\text{Im}\, T^{-1}(t)q\, dt$, and a quick calculation shows that

$$\text{Im}\, T^{-1}(t)q = \frac{q_2}{|A(t) - qC(t)|^2}, \quad q = q_1 + iq_2.$$

Now we turn this around. For $q \in \mathbb{C}^+$, define

$$M_q(z) = a_q + \frac{q_2}{\pi} \int_{-\infty}^{\infty} \left(\frac{1}{t-z} - \frac{t}{t^2+1} \right) \frac{dt}{|E_q(t)|^2}, \quad E_q = A - qC. \tag{4.19}$$

The choice of $a_q \in \mathbb{R}$ will be made later. The convergence of the integral will follow from the following estimate.

Lemma 4.21. *For $z \in \mathbb{C}^+ \cup \mathbb{R}$ and $q \in \mathbb{C}^+$, we have $|E(z)| \lesssim |E_q(z)| \lesssim |E(z)|$, and the implied constants depend only on q.*

We postpone the (easy) proof of this until after the proof of the theorem.

Notice that Lemma 4.21 does imply that

$$\int_{-\infty}^{\infty} \frac{dt}{(1+t^2)|E_q(t)|^2} < \infty,$$

as required in (4.19), because E has this property by the H^2 assumption. It is also useful to note here that $E \neq 0$ on $\mathbb{R}$, for the same reason, and thus A, C do not have common zeros.

So M_q is a Herglotz function, and I now claim that M_q has a holomorphic continuation to an open set $U_q \supseteq \mathbb{C}^+ \cup \mathbb{R}$. It suffices to show that for any $x \in \mathbb{R}$, M_q can be continued to some disk about x. To do this, observe that $|E_q(t)|^2 = E_q(t)E_q^{\#}(t)$ on $\mathbb{R}$, and this latter function is entire. Given $x \in \mathbb{R}$, fix a (closed) disk that does not contain any zeros of $E_q E_q^{\#}$, which is possible because there are no zeros on $\mathbb{R}$. Now deform the contour of integration in (4.19): replace the diameter of this disk by its lower semi-circle. This will not change (4.19) for $z \in \mathbb{C}^+$ (at least it will not if we avoided the folly of stepping into the poles $\pm i$ of the second term of the integrand), and now the modified representation makes it clear that $M_q(z)$ is holomorphic on our disk.

So in particular, $M_q(0)$ is well defined, and I now take a_q such that $\operatorname{Re} M_q(0) = q_1$. Since $\operatorname{Im} M_q(t) = q_2/|E_q(t)|^2$ for $t \in \mathbb{R}$, this then makes $M_q(0) = q$.

I now want to design functions B_q, D_q that make (4.18) happen for our M_q. This can be done systematically, by matching real and imaginary parts on the real line, but it is easier to just make the appropriate definitions without further excuse. We specialize to $q = i$, for the moment, and let

$$B(z) = -A(z)M_i(z) + \frac{i}{E(z)}, \quad D(z) = -C(z)M_i(z) + \frac{1}{E(z)}.$$

These formulae work for $z \in U_i \supseteq \mathbb{C}^+ \cup \mathbb{R}$ and define holomorphic functions there, but then we notice that B, D are real on the real line, so by the Schwarz reflection principle, $F^{\#}$ for $F = B, D$ provides a holomorphic extension to $\mathbb{C}^-$, and thus B, D are entire functions.

It is then straightforward to check that $B(0) = 0$, $D(0) = 1$, and $\det T(z) = 1$, with

$$T(z) = \begin{pmatrix} A(z) & B(z) \\ C(z) & D(z) \end{pmatrix}.$$

We now hope that $T \in TM$. The only property of functions in TM that has not been established yet for T is (4.14) or, equivalently, the Herglotz mapping property of $q \mapsto T^{-1}(z)q$, for $z \in \mathbb{C}^+$, so this is what we have to address now to finish the proof.

Define

$$g(q,z) = F_q(z) - M_q(z), \quad F_q(z) \equiv T^{-1}(z)q = -\frac{B(z) - qD(z)}{A(z) - qC(z)}.$$

Plugging the definitions of B, D into this, we obtain

$$F_q(z) = M_i(z) + \frac{q - i}{E(z)E_q(z)}.$$

By Lemma 4.21, $1/[(z + i)E_q(z)] \in H^2$ also, so

$$\frac{1}{(z + i)^2 E(z)E_q(z)} \in H^1.$$

This implies that $1/EE_q$ can be represented as the Poisson integral of its boundary values, as in (4.3). See Theorem 5.18 and Corollary 5.24 of [60] for more information on this fact. The imaginary parts of the Herglotz functions M_q, M_i are also equal to the

Poisson integrals of their boundary values; here it is important that we do not have a term bz in their definition (4.19), or else there would be an extra contribution of the form $b \operatorname{Im} z$. So $\operatorname{Im} g(q,z)$ is the Poisson integral of its boundary values, but since $\operatorname{Im} F_q(t) = q_2/|E_q(t)|^2$, by a calculation, which is the same as $\operatorname{Im} M_q(t)$, these are zero, and thus $g(q,z) = a(q)$ is constant. Evaluation at $z = 0$ then shows that $a(q) = 0$. So $F_q = M_q$, which is a Herglotz function, and thus

$$\operatorname{Im} F_q(z) = \operatorname{Im} T^{-1}(z)q > 0,$$

as we wished to show. □

Proof of Lemma 4.21. This is essentially the same argument that we already used at the end of the proof of Theorem 4.19. Recall that E being a de Branges function makes $-A/C$ a Herglotz function. So if we write

$$E_q = -iq_2 C\left(1 - i\frac{q_1}{q_2} + \frac{i}{q_2}\frac{A}{C}\right),$$

then it follows that $|E_q(z)| \geq q_2|C(z)|$ for $z \in \mathbb{C}^+$. This in turn implies that

$$\frac{E(z)}{E_q(z)} = 1 + \frac{(q-i)C(z)}{E_q(z)}$$

is bounded on $\mathbb{C}^+$, as required.

The proof of the second estimate is similar and in fact slightly easier still. □

Theorem 4.20 raises the question of whether E determines T, and clearly this is not the case since we can multiply from the right by a matrix with first column $(1,0)^t$ such as the special matrix not from TM_1 from Theorem 4.17. Note here that a product of matrices from TM will be in TM again. In fancier language, TM is a monoid (but certainly not a group: no $T \neq 1$ is invertible in TM).

It turns out that this is all the non-uniqueness here, and a somewhat stronger result along these lines is also true.

Theorem 4.22. *Let $T_1, T_2 \in TM$.*
(a) *If $A_1/B_1 = A_2/B_2$, then*

$$T_2(z) = \begin{pmatrix} 1 & 0 \\ az & 1 \end{pmatrix} T_1(z)$$

for some $a \in \mathbb{R}$.
(b) *Similarly, if $A_1/C_1 = A_2/C_2$, then*

$$T_2(z) = T_1(z)\begin{pmatrix} 1 & az \\ 0 & 1 \end{pmatrix}$$

for some $a \in \mathbb{R}$.

Of course, there are two more versions of this, which I have not made explicit. The quick summary is that any of the quotient Herglotz functions determines T, up to multiplication by one of the special matrices not in TM_1.

Proof. I will only discuss the case $T_1, T_2 \in TM_1$ explicitly. If one of the matrices is not in TM_1, then the argument I am going to give still works (and simplifies), but it would be inconvenient to constantly have to distinguish cases, so I leave it to the reader to make these adjustments.

(a) By assumption, $A_2(z) = F(z)A_1(z)$, $B_2(z) = F(z)B_1(z)$ for some function F, and since A_j, B_j do not have common zeros, it follows that F is entire and zero-free. By Theorem 4.19, $F \in N$, and since $F^{\#} = F$, Theorem 4.2 applies and shows that F is of exponential type. Since F does not have any zeros, its Hadamard product is simply $F(z) = e^{a+bz}$. Since $F(0) = 1$, we see that $a = 0$, and we also know that $b \in \mathbb{R}$, since $F = F^{\#}$. However, then $F(z) = e^{bz}$ will be in N only if $b = 0$.

We now know that $A_1 = A_2 =: A$, $B_1 = B_2 =: B$, and from the condition that both matrices have determinant 1, we can then conclude that

$$(C_2(z), D_2(z)) = (C_1(z), D_1(z)) + G(z)(A(z), B(z)).$$

As above, the fact that there are no common zeros of A, B implies that G is entire. We must show that $G(z) = az$.

To do this, we write G as a difference of Herglotz functions, as follows,

$$G(z) = \frac{C_2(z)}{A(z)} - \frac{C_1(z)}{A(z)}.$$

Both C_2/A and C_1/A are meromorphic and real on the real line, so their Herglotz representations are

$$\frac{C_j(z)}{A(z)} = a_j + b_j z + \sum_{E_n : A(E_n)=0} \left(\frac{1}{E_n - z} - \frac{E_n}{E_n^2 + 1} \right) g_n^{(j)}.$$

The difference of these can only be entire if it is of the form $G(z) = a + bz$, and taking $z = 0$, we see that $a = 0$ here.

The proof of part (b) is of course completely analogous. □

We close this section with one more interesting property of de Branges functions that come from a matrix $T \in TM$.

Theorem 4.23. *Let $T \in TM$ and, as usual, write $E = A - iC$. Then $|E(x + iy)|$ is an increasing function of $y \geq 0$ for every $x \in \mathbb{R}$.*

Proof. For any fixed $b > 0$, the functions $E_{\pm b}(z) = E(z \pm ib)$ will also lie in N. This is obvious for E_b, and for E_{-b}, it follows from criterion (4.4) because $|E(x-iy)| \leq |E(x+iy)|$ for $y > 0$ (the inequality is not strict here because we could have $T \notin TM_1$). Now recall the remarks about the canonical factorization of an *entire* function $F \in N$ that were made at the end of Section 4.1; see especially (4.7). When this is applied to the functions $E_{\pm b}(z)$, with $z = x + iy \in \mathbb{C}^+$, we obtain

$$\log|E(z+ib)| = \tau y + \frac{1}{\pi}\int_{-\infty}^{\infty} \frac{y}{(t-x)^2+y^2}\log|E(t+ib)|\,dt,$$

$$\log|E(z-ib)| \le \tau y + \frac{1}{\pi}\int_{-\infty}^{\infty} \frac{y}{(t-x)^2+y^2}\log|E(t-ib)|\,dt$$

$$\le \tau y + \frac{1}{\pi}\int_{-\infty}^{\infty} \frac{y}{(t-x)^2+y^2}\log|E(t+ib)|\,dt.$$

(The formula for E_{-b} is only an inequality, even before estimating $|E(t-ib)| \le |E(t+ib)|$, due to the possible presence of a Blaschke product; more precisely, the right-hand side is equal to $\log|E_{-b}(z)| - \log|B(z)|$, and $|B| \le 1$.) We have shown that $|E(z-ib)| \le |E(z+ib)|$, as asserted. □

4.5 Regular de Branges spaces

This section is an interlude. We elaborate on the theme of Theorems 4.19, 4.20. We will use the notation

$$(S_wF)(z) = \frac{F(z)-F(w)}{z-w};$$

if F is an entire function, then so is S_wF, for any $w \in \mathbb{C}$. The symbol S (as in shift) alludes to the fact that this operation shifts the Taylor coefficients of the expansion of F about w. This interpretation will play no role in what follows, though.

Theorem 4.24. *Let E be a de Branges function with $E(0) = 1$. Then the following statements are equivalent:*

(a) *$E = A - iC$ for some*

$$T(z) = \begin{pmatrix} A(z) & B(z) \\ C(z) & D(z) \end{pmatrix} \in TM;$$

(b)

$$\frac{1}{(z+i)E(z)} \in H^2;$$

(c) *for some $w \in \mathbb{C}$, we have $S_wF \in B(E)$ whenever $F \in B(E)$;*
(d) *for all $w \in \mathbb{C}$, we have $S_wF \in B(E)$ whenever $F \in B(E)$;*
(e) *$S_wE \in B(E)$ for some $w \in \mathbb{C}$ with $E(w) \ne 0$;*
(f) *$S_wE \in B(E)$ for all $w \in \mathbb{C}$.*

The condition that $E(0) = 1$ is a normalization and not essential. It can always be achieved by a change of de Branges function as in Theorem 4.5 if $E(0) \neq 0$.

The de Branges spaces satisfying the equivalent conditions of Theorem 4.24 are called *regular de Branges spaces.*

Proof. By Theorems 4.19, 4.20, (a) and (b) are equivalent. The implication from (b) to (d) is straightforward and follows from an argument very similar to the one in the first part of the proof of Theorem 4.4. We must check that $S_wF/E, (S_wF)^\#/E \in H^2$ if $F \in B(E)$, and the defining property of H^2 was given by (4.1). It is again clear that we need not worry about contributions coming from $|t+iy-w| \leq \delta$, with $\delta > 0$ fixed and sufficiently small, since either our function is bounded there or these points do not even lie in $\mathbb{C}^+$; here we use that E cannot have zeros on $\mathbb{C}^+$, being a de Branges function, and real zeros are prevented by our assumption that (b) holds.

Having dispensed of this part, we can then look separately at the two integrals

$$\int_{|t+iy-w|>\delta} \left| \frac{F(t+iy)}{(t+iy-w)E(t+iy)} \right|^2 dt,$$

$$\int_{|t+iy-w|>\delta} \left| \frac{F(w)}{(t+iy-w)E(t+iy)} \right|^2 dt,$$

and the supremum over $y > 0$ will indeed stay bounded for both, as required, since $F/E \in H^2$ and $1/[(z+i)E(z)] \in H^2$. The verification that $(S_wF)^\#/E \in H^2$ is of course completely analogous.

Next, we show that (c) implies (b). We start out by observing that there will be an $F \in B(E)$ with $F(w) \neq 0$. Indeed, if this was not true, then repeated application of our hypothesis would show that if $F \in B(E)$, then also $F/(z-w)^n \in B(E)$ for all $n \geq 1$, but this gives F a zero of infinite order at $z = w$, so $F \equiv 0$. However, $B(E)$ certainly contains non-zero elements, for example J_z for any $z \notin \mathbb{R}$, since then $J_z(z)$ is a non-zero multiple of $|E(z)|^2 - |E^\#(z)|^2$.

Take such an $F \in B(E)$, with $F(w) = 1$, say. We can then argue as above: consider

$$\frac{1}{(z-w)E(z)} = \frac{F(z)}{(z-w)E(z)} - \frac{(S_wF)(z)}{E(z)}.$$

Both terms on the right-hand side satisfy an H^2 type condition (4.1), but with the integrals again restricted to $|t+iy-w| \geq \delta$ for some $\delta > 0$. If $z = t+iy \in \mathbb{C}^+$ satisfies such a condition, then $z-w$ and $z+i$ become comparable in size, so $1/[(z+i)E(z)]$ satisfies the modified condition (4.1). We would now like to conclude that then this function also satisfies (4.1) itself, and the only potential problem here would be a zero of E at w, with $w \in \mathbb{R}$. This will not happen because $E(w) = 0$, $w \in \mathbb{R}$, would make $J_w \equiv 0$, which is impossible since there are functions F with $F(w) \neq 0$ in $B(E)$.

Finally, (e), (f) can be related to (b) in the same way. More explicitly, we deduce (f) from (b) by again showing that $S_w E/E$, $(S_w E)^\#/E \in H^2$, and the argument proceeds exactly as above.

If (e) is assumed, then we also argue as above, when we showed that (c) implies (b), and in fact the argument simplifies since now $E(w) \neq 0$ is part of the assumption and does not need to be shown separately. □

Corollary 4.25. *Assume that E satisfies the conditions of Theorem 4.24. Then the deficiency spaces of M_z are given by*

$$R(M_z - w)^\perp = L(J_w) \quad (w \in \mathbb{C}).$$

In Section 4.3, we studied M_z by using the fact that M_z is unitarily equivalent to the operator part of the relation $\mathcal{S}_0$, and we already identified the deficiency spaces there. The corollary provides an alternative argument for this part. The assumption on E is in fact unnecessary if $w \in \mathbb{C} \setminus \mathbb{R}$. Then the statement can be established in general, in much the same way as below, by using the fact that if $F \in B(E)$, $F(w) = 0$, then $\frac{z-\overline{w}}{z-w} F \in B(E)$ also, which is easy to show.

Proof. Obviously, if $G = (z - w)F \in R(M_z - w)$, then $[J_w, G] = G(w) = 0$, so $J_w \in R(M_z - w)^\perp$. Conversely, if $F \in B(E)$ has a zero at w, then, by Theorem 4.24(d), $F(z)/(z - w) \in B(E)$ also, so $F \in R(M_z - w)$. So $J_w^\perp \subseteq R(M_z - w)$ and thus $R(M_z - w)^\perp \subseteq L(J_w)$. □

4.6 The type of a transfer matrix

If $T \in TM$ is the transfer matrix of a canonical system, then there is a beautiful formula that computes $\tau = \tau(F)$, $F = A, B, C, D, E$, quite explicitly in terms of the coefficient function $H(x)$.

Theorem 4.26. *Consider a canonical system on $[0, L]$, with $H \in L^1(0, L)$. Then the type τ of $T(z) = T(L, 0; z) \in TM$ is given by*

$$\tau = \int_0^L \sqrt{\det H(x)}\, dx.$$

It is very easy to confirm that $\sqrt{\det H(x)} \in L^1$ if $H \in L^1$, so the integral is well defined.

One tool for the proof will be Gronwall's inequality; the following version will be sufficient for us here.

Lemma 4.27 (Gronwall). *Suppose that f, b are non-negative functions on $0 \leq x \leq L$, with f continuous and $b \in L^1(0, L)$, and*

$$f(x) \leq a + \int_0^x b(t) f(t)\, dt.$$

Then $f(x) \leq a e^{B(x)}$, with $B(x) = \int_0^x b(t)\, dt$.

Proof. Let $g(x) = e^{-B(x)} \int_0^x b(t)f(t)\,dt$. Then $g \in AC$, $g(0) = 0$,

$$g'(x) = -b(x)e^{-B(x)} \int_0^x b(t)f(t)\,dt + e^{-B(x)}b(x)f(x) \le ab(x)e^{-B(x)},$$

so $g(x) \le a \int_0^x b(t)e^{-B(t)}\,dt = a(1 - e^{-B(x)})$. This implies that

$$f(x) \le a + e^{B(x)}g(x) \le ae^{B(x)},$$

as claimed. □

Proof of Theorem 4.26. Denote the (mean) type of $T(x;z)$ by $\tau(x)$ and write $d(x) = \int_0^x \sqrt{\det H(t)}\,dt$. If $P \in \mathbb{C}^{2\times 2}$ is any invertible matrix, then

$$PT(x;z) = P + z\int_0^x (PJH(t)P^{-1})PT(t;z)\,dt,$$

so if we take (operator) norms and apply Gronwall's lemma to the resulting integral inequality, then we obtain

$$\|PT(x;z)\| \le \|P\| \exp\left(|z| \int_0^x \|PJH(t)P^{-1}\|\,dt \right).$$

Since $\|T\| \le \|P^{-1}\|\,\|PT\|$, this says that

$$\tau(x) \le \int_0^x \|PJH(t)P^{-1}\|\,dt, \tag{4.20}$$

for any P.

Denote the type of $T(y,x;z)$ similarly by $\tau(y,x)$. Of course, we can repeat the argument that led to (4.20) on the interval (x,y), and thus $\tau(y,x)$ obeys a similar bound. Since $T(y;z) = T(y,x;z)T(x;z)$, we have $\tau(y) \le \tau(x) + \tau(y,x)$. Moreover, $T(x,y;z) = T(y,x;z)^{-1}$ has the same entries as $T(y,x;z)$, except for possible signs, so $\tau(x,y) = \tau(y,x)$, and it thus follows that

$$|\tau(y) - \tau(x)| \le \tau(y,x) \le \left| \int_x^y \|PJH(t)P^{-1}\|\,dt \right|.$$

This implies that $\tau(x)$ is an absolutely continuous function, and

$$\tau'(x) \le \|PJH(x)P^{-1}\| \tag{4.21}$$

for almost every $x \in (0, L)$. This is true for any P; of course, we must keep in mind here that the null set on which the inequality does not hold can depend on P. However, we can choose a countable dense (in $\mathbb{C}^{2\times 2}$) subset $\{P_n\}$, and since the right-hand side of (4.21) is a continuous function of P, it does follow that also

$$\tau'(x) \le \inf_P \|PJH(x)P^{-1}\|$$

for almost every $x \in (0, L)$.

It is now easy to see that $\inf_P \|PJH(x)P^{-1}\| \le \sqrt{\det H(x)}$. Indeed, $\operatorname{tr} JH(x) = 0$, $\det JH(x) = \det H(x)$, so $JH(x)$ has eigenvalues $\pm i\sqrt{\det H(x)}$. If now also $\det H(x) > 0$, then these are distinct and we can take a P that diagonalizes JH to confirm the asserted bound. If $\det H(x) = 0$, then an approximation argument that refers to the case just discussed establishes the claim in this case also (or a direct argument making use of the Jordan normal form of JH again would of course work, too).

We have shown that $\tau'(x) \le d'(x)$ almost everywhere, and since $\tau(0) = d(0) = 0$, this implies that $\tau(x) \le d(x)$.

To prove the reverse inequality, we consider the modified transfer matrix $T_0(x; z) = e^{id(x)z}T(x; z)$. This solves $T_0(0; z) = 1$,

$$JT_0'(x) = -z(H(x) - i\sqrt{\det H(x)}J)T_0(x) = -zH_0(x)T_0(x),$$

with $H_0 = H - i\sqrt{\det H}J$. It is easy to see that $H_0(x) \ge 0$ also. The mean type τ_0 of T_0 satisfies $\tau_0(x) = \tau(x) - d(x)$; this is obvious if you use the formula $\tau = \limsup (1/y) \log |F(iy)|$ for the mean type of a function $F \in N$.

I would now like to refer to Theorem 4.19 to conclude that $\tau_0(x) \ge 0$, so $\tau(x) \ge d(x)$, as desired, but here we must be careful because H_0 has complex entries (and thus T_0 need not be real for real z), so our previous results do not apply literally. However, it is easy to verify that the arguments that led to the statement that $\tau_0 \ge 0$ still work, without any changes: first of all, we repeat the calculation from the proof of Theorem 1.2 to see that still $i(T_0^*(z)JT_0(z) - J) \ge 0$ for $z \in \mathbb{C}^+$; the crucial ingredient to this is the positivity of H_0. Then we repeat the argument from the first part of the proof of Theorem 4.19. □

4.7 Transfer matrices and singular intervals

If $T(z) \in TM$ is the transfer matrix $T(z) = T(L; z)$ of a canonical system $H \in L^1(0, L)$ and $(0, L)$ starts with a singular interval $(0, d)$ with specifically $H = hP_{e_2}$ there, then

$$T(z) = T(L, d; z)\begin{pmatrix} 1 & -az \\ 0 & 1 \end{pmatrix}, \quad a = \int_0^d h(x)\, dx > 0. \tag{4.22}$$

Of course, $T(L, d; z)$ also lies in TM, being a transfer matrix itself. We now prove that the converse of this observation also holds: if T can be factored in this way as the product of two matrices in TM, with the second one given by the special matrix above, then $(0, L)$ starts with a singular interval of type e_2. Later, we will recognize this result as a consequence of a much more general and powerful theorem, the correspondence between canonical systems and matrix functions $T \in TM$; see Theorem 5.2. That does not mean that our efforts here will be wasted; quite on the contrary, what we do here will become an important tool in the proof of this general result.

Let us introduce the notation

$$S_a(z) = \begin{pmatrix} 1 & -az \\ 0 & 1 \end{pmatrix}$$

for this special matrix function. Notice that $S_{-a} = S_a^{-1}$, and $S_a \in TM$ if and only if $a \geq 0$. For $T \in TM$, we then define

$$a(T) = \sup\{a \geq 0 : T(z)S_{-a}(z) \in TM\},$$

that is, we search for the largest possible a that allows a factorization as in (4.22).

Lemma 4.28. *For any $T \in TM$, we have $a(T) < \infty$ and $TS_{-a(T)} \in TM$.*

Proof. By specializing (4.14) to $z = iy$, with $y > 0$, and dividing by y, we obtain

$$\frac{T^*(iy)JT(iy) - J}{iy} \leq 0.$$

Since $T^*(iy) = T^t(-iy)$, these are difference quotients for $f(z) = T^t(-z)JT(z)$, so letting $y \to 0+$ shows that $f'(0) = -T^{t\prime}(0)J + JT'(0) \leq 0$. It is now easy to see that for the matrix function TS_{-a}, this condition is violated for all large a for any given T.

The second statement is obvious since the conditions defining TM are all preserved under taking limits. □

We first establish a characterization of $a(T)$ that applies to general abstract matrix functions $T \in TM$. For a Herglotz function M, we define

$$b(M) = \lim_{y\to\infty} \frac{M(iy)}{iy}.$$

The limit exists and $b(M) \geq 0$. Of course, this is nothing but the constant in the linear in z term $b(M)z$ from the Herglotz representation of M, or, if we use the version that was equation (3.9) of Chapter 3, then $b(M) = \nu(\{\infty\})$. To ensure formal correctness of Theorem 4.31, we also let $b(M) = 0$ if $M \equiv a \in \mathbb{R}_\infty$, so that $b(M)$ is now defined for all generalized Herglotz functions.

As a final preparation, recall that $T \in TM$ gives us Weyl disks

$$\mathcal{D}(z) = \{T^{-1}(z)q : q \in \overline{\mathbb{C}^+}\}.$$

Let us collect some basic facts about these.

Lemma 4.29. *If $C \not\equiv 0$, then for every $z \in \mathbb{C}^+$, $\mathcal{D}(z)$ is a disk of radius $1/R(z) = 2\,\mathrm{Im}\,\overline{A(z)}C(z) > 0$, and $\mathcal{D}(z) \subseteq \overline{\mathbb{C}^+}$.*

If $C \equiv 0$, then $T(z) = \begin{pmatrix} 1 & -az \\ 0 & 1 \end{pmatrix}$, and again everything is clear: then $\mathcal{D}(z) = \{w \in \mathbb{C} : \mathrm{Im}\, w \geq a\,\mathrm{Im}\, z\}$.

Proof. This is of course very similar to what we did earlier, and the Lemma is indeed more a reminder than a new result. The formula for the radius is Lemma 3.11 (and we did choose the correct sign), and if also $T \in TM_1$, then $\mathrm{Im}\,\overline{A}C = |A|^2 \mathrm{Im}\,(C/A) > 0$ on $\mathbb{C}^+$, by Theorem 4.18. This is also (trivially) true if $T \notin TM_1$, but $C \not\equiv 0$, by Theorem 4.17.

Condition (4.14) shows that $\mathcal{D}(z) \subseteq \overline{\mathbb{C}^+}$ (one could say that was its real meaning). □

Theorem 4.30. *If $C \not\equiv 0$, then $R(iy) \lesssim 1/y$ as $y \to \infty$.*

Proof. Since $C \not\equiv 0$, we know that $-A/C$ is a (genuine) Herglotz function. Any Herglotz function M satisfies $\mathrm{Im}\, M(iy) \gtrsim 1/y$ as $y \to \infty$, so

$$\frac{1}{R(iy)} = 2|C(iy)|^2\,\mathrm{Im}\left(-\frac{A(iy)}{C(iy)}\right) \gtrsim \frac{|C(iy)|^2}{y}.$$

Here C is of exponential type, by Theorem 4.19, its zeros c_n are real, $C(0) = 0$, and $C = C^\#$. Thus its Hadamard factorization reads

$$C(z) = ze^{a+bz} \prod_n \left(1 - \frac{z}{c_n}\right) e^{z/c_n},$$

with $a, b \in \mathbb{R}$ (I have also incorporated the fact that there are no multiple zeros, but this is not essential here and the argument would also work with a factor z^m instead of z). In particular,

$$|C(iy)|^2 = y^2 e^{2a} \prod_n \left(1 + \frac{y^2}{c_n^2}\right) \geq e^{2a} y^2,$$

and thus $1/R \gtrsim y$, as claimed. □

Theorem 4.31. *Let $T \in TM$. Then*

(a) *$a(T) = b(-B/A)$;*

(b) *if $C \not\equiv 0$, then for any generalized Herglotz function F with $F(z) \in \mathcal{D}(z)$ for all $z \in \mathbb{C}^+$, we have $b(F) = b(-B/A)$.*

These generalized Herglotz functions that lie in all Weyl disks, from part (b), have a convenient description: they are parametrized by the set of all generalized Herglotz functions q, by letting $F_q(z) = T^{-1}(z)q(z)$. Indeed, it is clear that such an F_q has the required properties. Conversely, if $F \in \mathcal{D}(z)$ for all $z \in \mathbb{C}^+$, then certainly $F(z) = T^{-1}(z)q(z)$ for some function q taking values in $\overline{\mathbb{C}^+}$, but then $q = TF$ will also be holomorphic.

Proof. We start with part (b), which will be used in the proof of part (a). We have $|F(z) + B(z)/A(z)| \le 2R(z)$ for any F that stays inside the Weyl disks, so the claim is an immediate consequence of Theorem 4.30.

(a) If $T = T_0 S_a$ with $T_0 \in TM$, then

$$-\frac{B}{A} = T^{-1}0 = S_{-a}T_0^{-1}0 = T_0^{-1}0 + az,$$

and since $T_0^{-1}0$ also is a generalized Herglotz function and $T_0^{-1}0 \not\equiv \infty$ (since $A \not\equiv 0$), we see that $b(-B/A) \ge a$. Hence $b(-B/A) \ge a(T)$.

As for the opposite inequality, we first observe that everything becomes trivial if $C \equiv 0$. In this case, $T = \left(\begin{smallmatrix} 1 & -az \\ 0 & 1 \end{smallmatrix}\right)$, so $-B/A = az$, and thus clearly $a(T) = a = b(-B/A)$. So we may now assume that $C \not\equiv 0$.

For any $q \in \overline{\mathbb{C}^+}$, the already established part (b) then shows that

$$T^{-1}(z)q = b(-B/A)z + F_q(z),$$

and here F_q also satisfies $\operatorname{Im} F_q \ge 0$ on $\mathbb{C}^+$. In other words,

$$\operatorname{Im}\left(S_{b(-B/A)}(z)T^{-1}(z)q\right) \ge 0$$

for all $q, z \in \mathbb{C}^+$, and this says that TS_{-b} satisfies (4.14), so $TS_{-b} \in TM$, and hence $a(T) \ge b(-B/A)$. □

We are ready to make the connection to the singular intervals, if now $T(z) = T(L;z)$ is the transfer matrix of a canonical system on $[0, L]$ with coefficient function $H \in L^1(0, L)$. In this case, we introduce the additional quantity

$$c(H) = \int_0^d h(x)\,dx,$$

and here $(0, d)$ denotes the initial singular interval with $H = hP_{e_2}$, if there indeed is one; if not, then we of course define $c(H) = 0$.

Theorem 4.32. *Let $T(z) = T(L;z) \in TM$ be the transfer matrix of a canonical system. Then $a(T) = c(H)$.*

Proof. The motivating discussion that opened this section has already established that $a(T) \ge c(H)$. To prove the opposite inequality, let us first observe that $C(x;z) = u_2(x,z) \not\equiv 0$ as a function of z for any $x > d$. Indeed, if we had $u_2 \equiv 0$, then every $z \in \mathbb{C}$ would become an eigenvalue of the problem on $[0, x]$ with boundary condition $u_2(x) = 0$, and this absurd situation can only be avoided if $(0, x)$ is a single singular interval. Moreover, the type of this singular interval would have to be specifically e_2 to make $u_2 \equiv 0$, but this now contradicts the choice of $x > d$.

So Theorem 4.31(b) applies to $T(x;z)$ for any such $x > d$, and we can define $b(x)$ as the common $b(F)$ of all generalized Herglotz functions F with $F(z) \in \mathcal{D}(x;z)$. However, these Weyl disks are also nested as x increases, so the Fs that may be used to compute $b(x_2)$ are a subset of those that determine $b(x_1)$ if $x_2 \ge x_1 > d$. So $b(x) = b(L) = b(-B/A)$ is constant on $(d, L]$. Now Theorem 4.31(a), applied to $T(x;z)$, shows that

$$T(x;z)S_{-a(T)} \in TM$$

for all $x > d$ and thus also $T(d;z)S_{-a(T)} \in TM$, by passing to the limit $x \to d+$. However, $T(d;z) = S_{c(H)}$, so $T(d;z)S_{-a(T)} = S_{c(H)-a(T)}$, which is in TM if and only if $c(H)-a(T) \ge 0$. □

In the next chapter, this equality $a = c$ will become an important tool. The additional statement from Theorem 4.31, that these are also equal to b, would then seem to be a stepping stone on the way to Theorem 4.32. However, these statements are also quite interesting in themselves, especially when applied to half line problems. They show how the large $|z|$ asymptotics of $m(z)$ correspond to what H does initially, close to $x = 0$.

Theorem 4.33. *Consider a half line problem on $[0,\infty)$, with limit point case at infinity, and let $m(z)$ be its m function. Then $b(m) > 0$ if and only if $(0,\infty)$ starts with a singular interval $(0,d)$ of type e_2, and in this case, if $H = hP_{e_2}$, then $b(m) = \int_0^d h(x)\,dx$.*

Proof. Of course, $m(z) \in \mathcal{D}(L;z)$ lies in all Weyl disks. So this just repeats (part of) what Theorems 4.31, 4.32 have to say about b. Recall that $C(x;z) = u_2(x;z) \not\equiv 0$ once we are past a possible initial singular interval, as we showed in the proof of Theorem 4.32. □

We can elaborate some more on this theme. Recall that in general, the measure ρ associated with $m(z)$ need not be finite; only $\int \frac{d\rho(t)}{1+t^2} < \infty$ is guaranteed. Theorem 4.33 now leads to the following satisfying criterion. A more general version will be proved in Section 6.2 as Theorem 6.9 there; for this, we will need additional tools.

Theorem 4.34. *Assume limit point case at infinity and let ρ be the measure associated with the m function $m(z)$. Then $\rho(\mathbb{R}) < \infty$ if and only if:*
(i) *$(0,\infty)$ starts with a singular interval of type $e_\alpha \neq e_2$, or*
(ii) *$(0,\infty)$ starts with a singular interval of type e_2, immediately followed by a second singular interval.*

We prepare for the proof by relating the condition that $\rho(\mathbb{R}) < \infty$ to the asymptotics of $m(z)$.

Lemma 4.35. *Let F be a Herglotz function with associated measure ρ. Then $\rho(\mathbb{R}) < \infty$ if and only if $F(iy) = a + iby + O(1/y)$ as $y \to \infty$ for some $a \in \mathbb{R}$, $b \ge 0$.*

Proof. If $\rho(\mathbb{R}) < \infty$, then the integral of $t/(t^2+1)$ may be absorbed by the constant a, so the Herglotz representation can be written in the form

$$F(z) = a + bz + \int_{-\infty}^{\infty} \frac{d\rho(t)}{t-z}.$$

It is now obvious that $F(iy) - a - iby = O(1/y)$, with these constants a, b.

Conversely, if this condition holds, then only $b = b(F)$ can work, since all other terms from the Herglotz representation are $o(y)$. So we may as well assume that $b = b(F) = 0$, and then, taking imaginary parts, we see that

$$y \operatorname{Im} F(iy) = \int_{-\infty}^{\infty} \frac{y^2}{t^2+y^2}\, d\rho(t) = O(1), \quad y \to \infty.$$

The integral converges to $\rho(\mathbb{R})$, by monotone convergence, so this finishes the proof. □

Proof of Theorem 4.34. We use the transformations

$$H_\alpha(x) = R_\alpha H(x) R_{-\alpha}, \quad R_\alpha = \begin{pmatrix} \cos\alpha & -\sin\alpha \\ \sin\alpha & \cos\alpha \end{pmatrix}.$$

By Theorem 3.20, $m_\alpha = R_\alpha m$. It is easy to show, by writing out the linear fractional transformation, that for any Herglotz function F and any $\alpha \neq 0$, we have $b(F) > 0$ if and only if $R_\alpha F(iy) = \cot\alpha + O(1/y)$ as $y \to \infty$. Now notice that H starts with a singular interval of type $\alpha^\perp$ if and only if H_α starts with a singular interval of type e_2. So, putting these observations together, we see that case (i) of the condition is equivalent to $m(iy) = a + O(1/y)$ as $y \to \infty$, for some $a \in \mathbb{R}$. By Lemma 4.35, (i) is therefore equivalent to the two conditions $\rho(\mathbb{R}) < \infty$, $b(m) = 0$.

We already know that $b(m) > 0$ if and only if $(0, L)$ starts with a singular interval $(0, d)$ of type e_2, and in this case, $m(z) = bz + m_d(z)$, with m_d denoting the m function of the problem on $[d, \infty)$, with boundary condition $u_2(d) = 0$. Obviously, (d, ∞) does not start with a singular interval of type e_2, and $b(m_d) = 0$, so we are now back to the case (i) just discussed and the Theorem follows. □

There is a mirror version of Theorem 4.32 which detects singular intervals of type e_1 at the right end of the basic interval $(0, L)$, and let me say a few words about this also. Given a canonical system on $[0, L]$, consider again its transfer matrix $T(z) = T(L; z) \in TM$ and let a_1, c_1 now refer to the right end of the interval, as follows: define

$$a_1(T) = \sup\left\{a \geq 0 : \begin{pmatrix} 1 & 0 \\ -az & 1 \end{pmatrix} T(z) \in TM\right\},$$

$$c_1(H) = \int_s^L h(x)\, dx,$$

and here (s, L) denotes the singular interval of type e_1 at the end of $(0, L)$ if there indeed is one; if not, then again $c_1(H) = 0$.

Theorem 4.36. *We have $a_1(T) = c_1(H)$.*

Proof. This is sufficiently similar to what we just did so that a sketch will suffice. One could adapt the whole discussion that led to Theorem 4.32 to this new setting (and then also relate a_1, c_1 to the linear in z term of certain Herglotz functions), but it is even easier to use the transformation

$$H_1(x) = -JH(L-x)J.$$

Since $-J = J^*$, this also satisfies $H_1(x) \geq 0$, so it is an admissible coefficient function, and its transfer matrix is given by

$$T_1(x;z) = -JT(L-x;-z)T^{-1}(L;-z)J,$$

so, in particular, $T_1(L;z) = -JT^{-1}(L;-z)J$. Now the theorem can be obtained by applying Theorem 4.32 to T_1. □

4.8 Notes

Section 4.1. The properties of the indicator function depend on F having exponential type; there are non-constant entire functions (not of finite order) that converge to zero along every ray.

The decomposition of ν from the canonical factorization (4.6) into singular and absolutely continuous parts corresponds to the factorization of F into what are called inner and outer functions.

Section 4.3. These results can, of course, all be found in [17], but there they are not, as here, approached from the point of view of spectral theory. I feel that this immediate connection to the canonical systems makes them much more intuitive and meaningful.

An interesting structural result that I do not present here is Theorem 23 of [17], which describes de Branges spaces as those spaces of entire functions where M_z has the properties we established: M_z is closed and symmetric, has deficiency indices 1, and is real with respect to the conjugation $F \mapsto F^{\#}$. In fact, this interpretation of the theorem is not immediately obvious (de Branges does point it out in the endnotes), and it has been missed by various authors who quoted it, including myself in [53].

Section 4.4. The method I use in the proof of Theorem 4.20 to reconstruct the missing second column of what we hope will eventually be a matrix function from TM could be new, even though it seems quite natural from a spectral theory point of view. In [17], de Branges uses a different tool to construct B, D that actually plays a rather large role in his work generally speaking: a map between different de Branges spaces that corresponds to a change of boundary condition at $x = 0$. In more concrete terms, if we had a canonical system, then we would send $F(z) = \int_0^L u^* Hf\, dx \in B(A - iC)$ to

$\widetilde{F}(z) = \int_0^L v^* Hf \, dx \in B(D + iB)$; this map can be constructed and studied (though quite laboriously) at the level of the de Branges spaces only.

The monotonicity of $y \mapsto |E(x + iy)|$ is studied in some detail in [17], in a more general context. The quick proof of Theorem 4.23 that I give here follows a hint to an exercise from [23]. It is also true that $y \mapsto |F(x + iy)|$, $y \geq 0$, is increasing for $F = A, B, C, D$. This is an immediate consequence of the Hadamard factorization of these functions, by a calculation similar to the one from the proof of Theorem 4.30.

Section 4.6. Theorem 4.26 is due to de Branges, and I closely follow his proof from [14] here; see also [59].

Section 4.7. The results on the characterization of $a(T)$ are due to de Branges, though they are not stated very explicitly in his work. In particular, the clever use of Weyl theory here to make the connection between $a(T)$ and $c(H)$ (via $b(-B/A)$) and Theorem 4.30 are due to de Branges. This technique will also play a crucial role in the next chapter, in the proof of the uniqueness part of Theorem 5.1.

Theorem 4.34 is from [69]. Strictly speaking, this is not the right place in this book to explore this topic, and I already announced that I will prove a more general result in Section 6.2, when more powerful tools will be available, but I could not resist the temptation to also receive some immediate payoff for our work here.

5 Inverse spectral theory

5.1 Introduction

In this chapter, we establish two related results that set up a one-to-one correspondence between certain spectral data and trace normed canonical systems. So from now on, $H(x)$ will be assumed to satisfy $\operatorname{tr} H(x) = 1$. Recall that for an arbitrary H on $[0, \infty)$, with 0 regular and limit point case at infinity, we can use the transformation $X = \int_0^x \operatorname{tr} H(t)\, dt$, $H_1(X) = H(x)/\operatorname{tr} H(x)$ to pass to an equivalent trace normed system on $X \in [0, \infty)$. The new system is equivalent in the sense that the transformation takes solutions to solutions in both directions and preserves norms, and thus also $m_1(z) = m(z)$. The change of variable also induces a unitary map $L^2_H \to L^2_{H_1}$ that transforms the corresponding relations (minimal, maximal, self-adjoint) into each other.

So it is really sufficient to consider trace normed systems, and we in fact have to do something of this sort if we want to obtain *unique* canonical systems for given spectral data.

In the formulation of the main results, we also drop our assumption that $(0, \infty)$ does not consist of a single singular interval; $H = P_\alpha$ on $(0, \infty)$ is now allowed. In part, this is done for reasons of formal elegance, but the main advantage is that our spaces become compact if these canonical systems are included. See the next section for a detailed discussion of this.

We will then need the m function also for this case $H(x) = P_\alpha$ on $(0, \infty)$ that was excluded from our original treatment in Chapter 3, but that poses no problems and was already discussed briefly in Section 3.7. We can again define $m(z) = f(0, z)$, with $Jf' = -zHf$, $f \in L^2_H(0, \infty)$. This solution is given by $f(x, z) = e_{\alpha^\perp}$, so $m(z) = -\tan\alpha \in \mathbb{R}_\infty$.

We still have the nested Weyl disks available, as could be established by revisiting our original treatment. But of course it is even easier to just work out everything explicitly: we have $T(L; z) = 1 + zLJP_\alpha$, and $\mathcal{D}(L; z) = T^{-1}(L; z)\overline{\mathbb{C}^+}$, so this is a disk of radius $R = 1/(2yL\cos^2\alpha)$ (writing $z = x + iy$, as usual) and with center $-\tan\alpha + iR$ if $\alpha \neq \pi/2$. (The only difference to the non-degenerate case discussed in Chapter 3 is that now $\mathcal{D}(L; z)$ touches the real line, as it must since $m(z) \in \mathbb{R}$.) If $\alpha = \pi/2$ or, equivalently, $H = P_{e_2}$, then $m(z) = \infty$, and $\mathcal{D}(L; z)$ is the half plane $\{w : \operatorname{Im} w \geq Ly\}$.

Theorem 5.1. *For every generalized Herglotz function F, there is a unique trace normed canonical system H on $[0, \infty)$ such that $m(z; H) = F(z)$.*

There is a similar one-to-one correspondence between trace normed canonical systems on bounded intervals and matrix functions $T \in TM$: given a $T \in TM$, there are unique $L \geq 0$ and $H \in L^1(0, L)$ such that $T(z) = T(L; z)$ is the transfer matrix of this canonical system.

https://doi.org/10.1515/9783110563238-005

The length L of the interval can be determined right away. Since our canonical systems are trace normed now, we have $L = \int_0^L \operatorname{tr} H(x)\, dx$, and since $JT_z'(x; z = 0) = -H(x)$, $T_z(0; 0) = 0$ (writing $T_z \equiv \partial T/\partial z$), this equals $-\operatorname{tr} JT_z(z = 0) = C'(0) - B'(0)$.

Motivated by this, let us introduce

$$TM(L) = \{T \in TM : C'(0) - B'(0) = L\}.$$

Since $T^{t\prime}(0)J - JT'(0) \geq 0$ also for a general $T \in TM$, as we observed earlier, in the proof of Lemma 4.28, it follows by taking the trace of this matrix that $C'(0) - B'(0) \geq 0$ for any $T \in TM$. So

$$TM = \bigcup_{L \geq 0} TM(L),$$

and the union is disjoint.

Theorem 5.2. *Let $L \geq 0$. For every $T \in TM(L)$, there is a unique trace normed canonical system $H(x)$ on $[0, L]$ such that $T(z) = T(L; z; H)$.*

Let us introduce some notation for the spaces involved here. We will write $\mathcal{C}$ for the collection of (the coefficient functions of) trace normed canonical systems on $[0, \infty)$; similarly, $\mathcal{C}(L)$ will refer to the set of trace normed canonical systems on $[0, L]$. The space of generalized Herglotz functions will be denoted by $\mathcal{F}$.

Theorems 5.1, 5.2 then provide bijections $\mathcal{C} \to \mathcal{F}$, $\mathcal{C}(L) \to TM(L)$. More is true: both maps become homeomorphisms between compact metric spaces if $\mathcal{C}$, $\mathcal{C}(L)$, $\mathcal{F}$, $TM(L)$ are endowed with certain natural metrics. Please see Corollary 5.8 below for the formal statement.

This is of course quite interesting in its own right; in addition, these metrics will also play an important role in the proofs of Theorems 5.1, 5.2. We will discuss them in detail in the next section.

5.2 Metrics on canonical systems and spectral data

We begin with the space $\mathcal{C}(L)$ of trace normed coefficient functions $H(x)$ on $0 < x < L$. We identify H with the measure $H(x)\, dx$ and then put the weak-$*$ topology on this space, which is available since measures can be viewed as the dual of continuous functions. The desired compactness property will then be a consequence of the Banach–Alaoglu Theorem.

While these facts are helpful as background information, it is perfectly feasible to give a treatment that does not rely on any machinery, except for one reference to the Riesz representation theorem, and I will do this below. As a final preparatory remark, note that $H(x)\, dx$ is a matrix valued measure, not a positive measure of the traditional type: it assigns the positive definite matrix $\int_A H(x)\, dx$ to a Borel set $A \subseteq [0, L]$.

We actually wanted a metric, not just a topology, and we define it as follows. Fix a countable collection of continuous functions $f_n \in C[0, L]$ (but taking values in $\mathbb{C}^2$)

that is dense with respect to the uniform norm; for example, polynomials with rational coefficients as the component functions would work for this. For $H_1, H_2 \in \mathcal{C}(L)$, define

$$\rho_n(H_1, H_2) = \left| \int_0^L f_n^*(x)(H_1(x) - H_2(x)) f_n(x)\, dx \right|, \tag{5.1}$$

and then

$$d(H_1, H_2) = \sum_{n\geq 1} 2^{-n} \frac{\rho_n(H_1, H_2)}{1 + \rho_n(H_1, H_2)}.$$

This is a metric on $\mathcal{C}(L)$; the individual ρ_ns are not guaranteed to be metrics, though, since it could happen that $\rho_n(H_1, H_2) = 0$ for some $H_1 \neq H_2$. Clearly, $d(H_n, H) \to 0$ if and only if $\rho_j(H_n, H) \to 0$ for every $j \geq 1$.

Theorem 5.3. *Let $H_n, H \in \mathcal{C}(L)$. Then the following statements are equivalent:*
(a) $d(H_n, H) \to 0$;
(b) *for all continuous functions f, we have*

$$\int_0^L H_n(x) f(x)\, dx \to \int_0^L H(x) f(x)\, dx;$$

(c) *for all continuous functions f, we have*

$$\int_0^L f^*(x) H_n(x) f(x)\, dx \to \int_0^L f^*(x) H(x) f(x)\, dx;$$

(d) *for all continuous functions f, g, we have*

$$\int_0^L g^*(x) H_n(x) f(x)\, dx \to \int_0^L g^*(x) H(x) f(x)\, dx.$$

Proof. It is easy to see that (b), (c), (d) are equivalent: to deduce (d) from (c), use polarization, and nothing else poses any problems.

Next, (a) implies the condition from (c), initially for $f = f_j$, but then also for all f by approximation. Here it is important that $\|H(x)\| \leq 1$ for all x and for any $H \in \mathcal{C}(L)$, since $H(x) \geq 0$ and $\operatorname{tr} H(x) = 1$. Conversely, (c) for $f = f_j$ shows that $\rho_j(H_n, H) \to 0$ for all j, and this gives (a). □

Theorem 5.4. *The space $(\mathcal{C}(L), d)$ is compact.*

Proof. I already outlined a possible argument above, and it might be helpful to recall this overview one more time before plunging into the details. Since d generates the weak-∗ topology, a compactness property is built into it, for example by the Banach–Alaoglu theorem. More precisely, a compact space is obtained if one collects all measures satisfying a given uniform bound, and this would normally include singular measures. In our situation, the normalization of the trace will make sure that all limiting measures are in $\mathcal{C}(L)$ again.

This sketch will probably be completely convincing already if you have seen similar things before, but let me also provide a more detailed version that does not rely on machinery. We must show that every sequence $H_n \in \mathcal{C}(L)$ has a convergent subsequence. Since

$$\left| \int_0^L g^*(x) H_n(x) f(x)\, dx \right| \le L \|f\|_\infty \|g\|_\infty ,$$

we can certainly make the sequence of these integrals convergent on a subsequence for $f = f_j$, $g = f_k$, first for fixed j, k, but then also for all j, k by a diagonal process. We will denote this subsequence by H_n also, for notational convenience. Since the f_j are dense, a similar estimate then shows that the limit

$$I(g, f) \equiv \lim_{n\to\infty} \int_0^L g^*(x) H_n(x) f(x)\, dx$$

exists for all continuous functions f, g, and $|I(g, f)| \le L\|f\|\,\|g\|$. Here we can in particular take functions of the form $f(x) = s(x) e_j$, with s mapping to $\mathbb{C}$ (rather than $\mathbb{C}^2$) now, and thus the Riesz representation theorem provides (signed) finite regular Borel measures μ_{jk}, $j, k = 1, 2$, on $[0, L]$, such that

$$I(g, f) = \int_0^L g^*(x) d\mu(x) f(x), \quad \mu(A) = \begin{pmatrix} \mu_{11}(A) & \mu_{12}(A) \\ \mu_{12}(A) & \mu_{22}(A) \end{pmatrix} \ge 0.$$

We now want to show that $d\mu(x) = H(x)\, dx$, for some $H \in \mathcal{C}(L)$. If f is a continuous function with $\|f\| \le 1$ and $f(x) = 0$ for $x \notin U$ for some Borel set $U \subseteq [0, L]$, then clearly

$$\int_0^L f^*(x) H_n(x) f(x)\, dx \le |U|,$$

so $\int f^* d\mu f \le |U|$ also. By the regularity of μ, this implies that μ is absolutely continuous (approximate a set of measure zero from the outside by open sets).

So we can write $d\mu(x) = H(x)\, dx$ for some $H \in L^1(0, L)$, $H(x) \ge 0$. We must also show that $\operatorname{tr} H = 1$ here, but this is obvious from taking $f(x) = s(x) e_j$ and then passing to the limits of

$$\int_0^L f^*(x) H_n(x) f(x)\, dx = \int_0^L e_j^* H_n(x) e_j |s(x)|^2 dx$$

and taking the sum over $j = 1, 2$.

Finally, it is clear from the construction of H and Theorem 5.3 that $d(H_n, H) \to 0$ along the subsequence that was chosen earlier. □

Now the most convenient way to build an appropriate metric on $\mathcal{C}$ is to view this as a product of spaces $\mathcal{C}([n-1, n])$ and then give it its product topology. So let d_n be a metric as above, for trace normed coefficient functions $H \in L^1(n-1, n)$, and then, for $H_1, H_2 \in \mathcal{C}$, define

$$d(H_1, H_2) = \sum_{n\geq 1} 2^{-n} \frac{d_n(H_1^{(n)}, H_2^{(n)})}{1 + d_n(H_1^{(n)}, H_2^{(n)})};$$

here, $H^{(n)}$ denotes the restriction of H to $(n-1, n)$, as is already clear from the context.

One does not have to go through this decomposition of the basic interval $[0, \infty)$; an equivalent metric is obtained by running the above construction directly. More specifically, fix a countable uniformly dense subset of the space of compactly supported continuous ($\mathbb{C}^2$ valued) functions on $[0, \infty)$, define ρ_n as in (5.1), and let

$$d'(H_1, H_2) = \sum_{n\geq 1} 2^{-n} \frac{\rho_n(H_1, H_2)}{1 + \rho_n(H_1, H_2)}.$$

Our above results immediately give the following.

Theorem 5.5. (a) *Let $H_n, H \in \mathcal{C}$. Then $d(H_n, H) \to 0$ if and only if*

$$\int_0^\infty f^*(x)H_n(x)f(x)\,dx \to \int_0^\infty f^*(x)H(x)f(x)\,dx$$

for all continuous compactly supported functions f.
(b) *The space $(\mathcal{C}, d)$ is compact.*

On the set $\mathcal{F}$ of generalized Herglotz functions, we use the topology of locally uniform convergence. This topology is metrizable, and a straightforward way to obtain such a metric would be to exhaust $\mathbb{C}^+$ by compact sets $K_n \subseteq \mathbb{C}^+$, use these to define metrics $\rho_n(F, G) = \max_{z\in K_n} \delta(F(z), G(z))$, and then form a weighted sum as above. Here, δ denotes the spherical metric on $\mathbb{C}_\infty$, so we compare the values as points on the Riemann sphere. This precaution is necessary since we could have F or $G \equiv \infty$; also, we would not obtain the desired compactness property if we did not have a value that is close to all large values.

However, an even easier procedure would be to take just a single K with an accumulation point in $\mathbb{C}^+$; for example, we can take $K = \{z : |z - 2i| \leq 1\}$. Then the metric

$$d(F, G) = \max_{z\in K} \delta(F(z), G(z)) \tag{5.2}$$

also generates the topology of locally uniform convergence. The reason behind this fact, that one compact set K controls all the others, is the same as the one that leads to the compactness of $\mathcal{F}$, so let us discuss this first. Any sequence $F_n \in \mathcal{F}$ is a normal family, so existence of a locally uniformly convergent subsequence is indeed guaranteed, and its limit will be holomorphic and will map to $\overline{\mathbb{C}^+}$, so will lie in $\mathcal{F}$ itself, as required.

Now let us go back to (5.2). We must show that if $F_n \to F$ uniformly on K and $K' \subseteq \mathbb{C}^+$ is any compact subset, then also $F_n \to F$ uniformly on K'. If this were false, then by the normal families argument just presented we would have $F_n \to G$ on a subsequence (for convenience denoted by F_n again), for some $G \in \mathcal{F}$, $G \neq F$. This is impossible because G has to equal F on K, so $G = F$ everywhere.

Let us summarize.

Theorem 5.6. (a) *Let $F_n, F \in \mathcal{F}$. Then $d(F_n, F) \to 0$ if and only if $F_n(z) \to F(z)$ locally uniformly on $\mathbb{C}^+$.*
(b) *The space $(\mathcal{F}, d)$ is compact.*

Finally, on $TM(L)$ we do the same thing. We use the topology of locally uniform convergence, and again it is easy to introduce a metric d that generates this topology. Then $(TM(L), d)$ also is a compact space, but this time, it is not entirely straightforward to show this directly, and it is not necessary for us either since we will obtain the compactness of $(TM(L), d)$ as an automatic byproduct of Theorem 5.2, combined with Theorem 5.7(a) below.

Theorem 5.7. (a) *Let $H_n, H \in \mathcal{C}(L)$ and suppose that $d(H_n, H) \to 0$. Then $T(x; z; H_n) \to T(x; z; H)$ uniformly on $0 \le x \le L$, $z \in K$ for every compact subset $K \subseteq \mathbb{C}$.*
(b) *The map $\mathcal{C} \to \mathcal{F}$, $H \mapsto m(z)$ is continuous.*

Part (a) says the map $\mathcal{C}(L) \to TM(L)$, $H \mapsto T(L; z)$ is continuous, and it makes a slightly stronger statement along these lines. In particular, once we know from Theorem 5.2 that this map is onto, it will then follow, as promised, that $TM(L)$, being the image of a compact space under a continuous map, is compact itself.

Proof. (a) Let us first establish this convergence at a fixed $z \in \mathbb{C}$. Write T_n and T for the transfer matrices of H_n and H, respectively. Since $JT_n' = -zH_nT_n$, $T_n(0) = 1$, and $\|H_n(x)\| \le 1$ for a trace normed H_n, we have

$$\|T_n(x)\| \le 1 + |z| \int_0^x \|T_n(t)\|\, dt,$$

so Gronwall's Lemma 4.27 shows that $\|T_n(x)\| \le e^{L|z|}$, $0 \le x \le L$. Since $T_n' = zJH_nT_n$, this implies that $\{T_n\}$ is an equicontinuous sequence of functions on $[0, L]$. The Arzela–Ascoli theorem now gives us the existence of a uniformly convergent subsequence $T_n(x) \to S(x)$, which, for notational convenience, I have denoted by T_n again.

Then

$$\int_0^x H_n(t)T_n(t)\,dt = \int_0^x H_n(t)S(t)\,dt + \int_0^x H_n(t)(T_n(t) - S(t))\,dt$$

converges to $\int_0^x H(t)S(t)\,dt$ because the first term on the right-hand side has this limit, and the fact that H_n is trace normed gives us uniform control on the second one. Note also that while $\chi_{[0,x]}(t)S(t)$ is not a continuous test function and thus the convergence is not literally a consequence of Theorem 5.3, there are no real problems because we can replace the characteristic function by a continuous cut-off, and we again have uniform control on the error introduced by this from the condition that $\operatorname{tr} H_n = 1$.

Now we can pass to the limit in the canonical system, written in integral form as

$$T_n(x) = 1 + zJ\int_0^x H_n(t)T_n(t)\,dt,$$

to conclude that $S(x) = 1 + zJ\int_0^x H(t)S(t)\,dt$. In other words, $S(x) = T(x)$ is the transfer matrix of H.

This argument has also shown that $T(x)$ is the only possible limit of T_n on any subsequence, and this forces the original sequence to converge to $T(x)$ also, without the need to pass to a subsequence. Indeed, if this was not true, then the compactness argument based on the Arzela–Ascoli theorem would produce a subsequence $T_{n_j} \to S \neq T$, but we just saw that this is impossible.

We have established the convergence statement of part (a) at a fixed $z \in \mathbb{C}$, and it is now easy to upgrade this to locally uniform (in z) convergence. To do this, use Gronwall's lemma to also control $T_z = \partial T/\partial z$. This solves $JT_z' = -zHT_z - HT$, $T_z(0) = 0$. Pass to the integrated form, recall that $\|T\| \le e^{L|z|}$, and then refer to Lemma 4.27. This will show that $\|T_z(x;z)\| \le Le^{2L|z|}$ and of course $(T_n)_z$ obeys the same bound, so the convergence is indeed uniform in z on compact sets.

(b) This follows quickly from part (a). It suffices to show that if $d(H_n, H) \to 0$, then for any fixed $z \in \mathbb{C}^+$, we have $m_n(z) \to m(z)$. The compactness of $\mathcal{F}$ will then automatically upgrade this to locally uniform convergence.

We use Weyl theory. Let us first treat the case where H is not identically equal to P_{e_2} on $(0,\infty)$. Then the Weyl disks are genuine disks (eventually) that shrink to $m(z)$, so if we make $L > 0$ large enough, then all points of $\mathcal{D}(L;z) = T^{-1}(L;z)\overline{\mathbb{C}^+}$ will be close to $m(z)$. But then the same is also true for all points of $\mathcal{D}_n(L;z) = T_n^{-1}(L;z)\overline{\mathbb{C}^+}$ for all large n because then T_n^{-1} is close to T^{-1}, by part (a). To confirm that the disk is a continuous function of the matrix T^{-1}, recall that the boundary circle may be obtained as the set of points

$$T^{-1}(z)e_\alpha = \frac{D(z)\cos\alpha - B(z)\sin\alpha}{A(z)\sin\alpha - C(z)\cos\alpha}, \quad 0 \le \alpha < \pi,$$

and here the denominator is bounded away from zero, uniformly in α; if it was not, then $\mathcal{D}(L;z)$ would contain ∞ and thus not be a genuine disk in the plane. Thus small perturbations of A, B, C, D will indeed lead to a circle that is close.

If $H \equiv P_{e_2}$, so $m(z) \equiv \infty$, then this argument still works. Now

$$T^{-1}(L;z) = \begin{pmatrix} 1 & Lz \\ 0 & 1 \end{pmatrix},$$

we again take a large $L > 0$, and it is then still true that a matrix $T_n^{-1}(L;z)$ close to $T^{-1}(L;z)$ will have only large points in its Weyl disk. Perhaps the easiest way to see this would be to apply the conformal map $w \mapsto -1/w$ to move things from ∞ to 0, so consider $JT^{-1}\overline{\mathbb{C}^+}$, and then apply the argument from the previous paragraph to this. □

As an immediate consequence of Theorem 5.7, we now also obtain the promised sharpened version of Theorems 5.1, 5.2.

Corollary 5.8. *The maps $\mathcal{C} \to \mathcal{F}$, $H \mapsto m(z)$, and $\mathcal{C}(L) \to TM(L)$, $H \mapsto T(L;z)$ are homeomorphisms between compact metric spaces.*

Proof (assuming Theorems 5.1, 5.2). These maps are continuous by Theorem 5.7, and Theorems 5.1, 5.2 show that they are bijections. A continuous bijection between compact metric spaces automatically has a continuous inverse. □

5.3 Existence of a canonical system with given *m* function

In this section, we prove the existence part of Theorem 5.1. This will not be very difficult. We will show directly that a dense subset of $\mathcal{F}$ can be realized as the m functions of canonical systems, and then the general claim will follow from the compactness of $\mathcal{C}$ by approximation.

I offer two completely different approaches in this first step. The first one will be direct and hands-on. The second one will essentially consist of a reference to the well-known fact that any compactly supported probability measure can be realized as the spectral measure of a Jacobi difference operator, and then we will rewrite these as canonical systems.

Let us now carry out the first approach. The starting point is the observation that any function from $(\mathcal{F}, d)$ can be approximated by Herglotz functions M with finitely supported measures, of the form

$$M(z) = a + bz + \sum_{j=1}^{N} \frac{g_j}{E_j - z}.$$

This is obvious from the Herglotz representation theorem, by approximating the original measure by such finitely supported measures, and for this step, the modified representation with a finite measure on the compact space $\mathbb{R}_\infty$ is especially convenient.

Such an M is rational, with all poles and zeros on the real line, and can thus be written as the quotient of two polynomials $M(z) = F(z)/G(z)$ that have the following properties:

$$F = F^{\#}, \quad G = G^{\#}; \quad F, G \text{ have no common zeros; } \frac{F}{G} \in \mathcal{F}. \tag{5.3}$$

We will now show that such an M can be realized as the m function of a canonical system with singular intervals only, and the basic step is provided by the following result.

Lemma 5.9. *Let F, G be two polynomials satisfying* (5.3). *Suppose that* $n := \max\{\deg F, \deg G\} \ge 1$. *Then there are $L > 0$ and $\alpha \in [0, \pi)$ such that the new polynomials F_1, G_1 defined by*

$$\begin{pmatrix} F(z) \\ G(z) \end{pmatrix} = (1 - zLJP_\alpha) \begin{pmatrix} F_1(z) \\ G_1(z) \end{pmatrix}$$

still satisfy (5.3), *and* $\max\{\deg F_1, \deg G_1\} = n - 1$.

Note that $1 - zLJP_\alpha = T^{-1}(L; z)$ is the inverse of the transfer matrix across $(0, L)$ of the canonical system with $H = P_\alpha$ there.

Proof. It will be convenient to work with the transformed polynomials

$$\begin{pmatrix} F_\theta(z) \\ G_\theta(z) \end{pmatrix} := R_\theta \begin{pmatrix} F(z) \\ G(z) \end{pmatrix}, \quad R_\theta = \begin{pmatrix} \cos\theta & -\sin\theta \\ \sin\theta & \cos\theta \end{pmatrix}.$$

Observe, first of all, that for any given θ, the polynomials F, G will satisfy (5.3) if and only if F_θ, G_θ do. Next, I claim that it suffices to find L, α with the required properties for F_θ, G_θ for some θ. Indeed, once we have such L, α, which are then used to obtain F_1, G_1 from F_θ, G_θ, then it will also follow that

$$\begin{pmatrix} F(z) \\ G(z) \end{pmatrix} = (1 - zLR_{-\theta}JP_\alpha R_\theta) \begin{pmatrix} F_2(z) \\ G_2(z) \end{pmatrix}, \quad \begin{pmatrix} F_2(z) \\ G_2(z) \end{pmatrix} = R_{-\theta} \begin{pmatrix} F_1(z) \\ G_1(z) \end{pmatrix},$$

and since $R_{-\theta}JP_\alpha R_\theta = JP_{\alpha-\theta}$, this is what we set out to do.

We can write

$$\begin{pmatrix} F(z) \\ G(z) \end{pmatrix} = s_n z^n e_\beta + \text{lower-order terms,}$$

with $s_n \in \mathbb{R}$, $s_n \neq 0$, and by making use of the reduction we just discussed, we may assume that $e_\beta = e_1$. Then $F/G = bz + M$, with $b = b(F/G) > 0$ and $M \in \mathcal{F}$ also. It also follows that $\deg G = n - 1$, or otherwise we would obtain the asymptotics $F/G \simeq z^k$, $k \ge 2$, which are incompatible with the Herglotz property of F/G.

I now claim that $L = b$, $P_\alpha = P_{e_2}$ have the desired properties. With these choices, we have

$$\frac{F_1(z)}{G_1(z)} = (1 + zbJP_{e_2})\begin{pmatrix} F(z) \\ G(z) \end{pmatrix} = \frac{F(z)}{G(z)} - bz \in \mathcal{F},$$

as required. Furthermore, it is clear that $F_1 = F - bzG$, $G_1 = G$ satisfy the remaining requirements of (5.3), and $\deg G_1 = \deg G = n - 1$. Since $b(F_1/G_1) = 0$, we must then also have $\deg F_1 \le n - 1$. □

We can now finish the proof of the existence claim of Theorem 5.1 in a few lines. If $M = F/G \in \mathcal{F}$ is of the type just discussed, with finitely supported associated measure, then we apply Lemma 5.9 until the degree is zero, and this shows that

$$M(z) = (1 - zL_1JP_{\alpha_1}) \cdots (1 - zL_nJP_{\alpha_n})e_\theta = T^{-1}(L; z)e_\theta$$

for some θ; as I already mentioned above, the matrix product is the inverse of the transfer matrix of the canonical system that consists of a succession of singular intervals of lengths L_j and types α_j. If we now follow these up with a singular half line of type $\theta^\perp$, then we have produced a canonical system $H \in \mathcal{C}$ with $m(z; H) = M(z)$.

If a general $F \in \mathcal{F}$ is given, then we approximate it by functions $M_n \in \mathcal{F}$, $d(M_n, F) \to 0$, of the type just discussed. These are realized by canonical systems $H_n \in \mathcal{C}$, and now the compactness of $\mathcal{C}$ lets us extract a convergent subsequence $H_{n_j} \to H \in \mathcal{C}$. By Theorem 5.7(b), $M_{n_j} \to m(z; H)$, so $m(z; H) = F(z)$, as desired.

Let me now discuss the alternative approach that could replace the first part of the argument. This will make use of *Jacobi matrices* on a half line $\mathbb{Z}_+ = \{1, 2, \dots\}$. These are self-adjoint operators on $\ell^2(\mathbb{Z}_+)$ that are associated with the difference equation

$$a_n y_{n+1} + a_{n-1} y_{n-1} + b_n y_n = z y_n. \tag{5.4}$$

For our purposes here, it suffices to consider bounded coefficient sequences $a, b \in \ell^\infty$, $a_n > 0$, $b_n \in \mathbb{R}$. Then we automatically have limit point case at infinity, and the m function of the Jacobi matrix can be defined as

$$m_J(z) = -\frac{f_1}{a_0 f_0}, \quad f \text{ solves (5.4)}, f \in \ell^2(\mathbb{Z}_+).$$

As usual, it is advisable to use the transfer matrix formalism, and we introduce these as follows here. For a sequence y_n, let $Y_n = (y_{n+1}, -a_n y_n)^t$. Then y solves (5.4) if and only if Y solves

$$Y_n = (A_n + zB_n)Y_{n-1}, \quad A_n = \begin{pmatrix} -b_n/a_n & 1/a_n \\ -a_n & 0 \end{pmatrix}, \quad B_n = \frac{1}{a_n}\begin{pmatrix} 1 & 0 \\ 0 & 0 \end{pmatrix}.$$

The transfer matrix T is then defined as the matrix solution of this equation, with initial value $T = 1$. We will specifically need the transfer matrix T_0 at $z = 0$ here.

So this matrix is defined as the solution of $T_0(0) = 1$, $T_0(n) = A_n T_0(n-1)$. Its matrix elements are given by

$$T_0(n) = \begin{pmatrix} p_{n+1} & q_{n+1} \\ -a_n p_n & -a_n q_n \end{pmatrix};$$

the sequences p, q solve (5.4), and they have the initial values $p_1 = -a_0 q_0 = 1$, $q_1 = a_0 p_0 = 0$.

To rewrite the Jacobi equation as a canonical system, we proceed as in Section 1.3. Given a sequence y_n, introduce the new sequence u_n by writing $Y_n = T_0(n)u_n$ (variation of constants about $z = 0$). A calculation then shows that y solves (5.4) if and only if u solves

$$J(u_n - u_{n-1}) = -zH_n u_{n-1}, \quad H_n = \begin{pmatrix} p_n^2 & p_n q_n \\ p_n q_n & q_n^2 \end{pmatrix}. \tag{5.5}$$

As we already saw in Section 1.2, this recursion can be obtained from a genuine canonical system. If we want it to be trace normed, then we must define

$$H(x) = \frac{1}{L_n} H_n \quad (c_{n-1} < x < c_n),$$

and here $L_n = p_n^2 + q_n^2$, $c_0 = 0$, $c_n = c_{n-1} + L_n$. The connection between this canonical system and its difference version is that if $u(x)$ solves $Ju' = -zHu$, then the sequence $u_n := u(c_n)$ will solve (5.5).

We should now think of $Ju' = -zHu$ as a rewriting of the original Jacobi equation (5.4). For our purposes here in this section, the following aspect of this connection is the most important one.

Theorem 5.10. *The m function of the canonical system just constructed is the m function of the Jacobi operator: $m(z; H) = m_J(z)$.*

Proof. The transfer matrix across a singular interval is given by

$$T(x, c_{n-1}; z) = 1 + z\frac{x - c_{n-1}}{L_n} JH_n,$$

and $H_n J H_n = 0$, as we already noticed in Section 1.2. Equipped with these observations, we can now see from a calculation that if u solves $Ju' = -zHu$ and $Y_n = T_0(n)u(c_n)$, then

$$\int_{c_{n-1}}^{c_n} u^*(x)H(x)u(x)\, dx = \frac{1}{a_n^2} Y_n^* \begin{pmatrix} 0 & 0 \\ 0 & 1 \end{pmatrix} Y_n. \tag{5.6}$$

In particular, we can now take the ℓ^2 solution f of (5.4). Then $m_J(z) = F_0(z)$, with, as above, $F_n = (f_{n+1}, -a_n f_n)^t$. If u denotes the solution of $Ju' = -zHu$ with the same initial value $u(0, z) = F_0(z)$, then, since $T_0(0) = 1$, we have $T_0(n)u(c_n, z) = F_n(z)$. So (5.6) applied to this u says that $\int_{c_{n-1}}^{c_n} u^* Hu = |f_n|^2$. Thus $u \in L_H^2(0, \infty)$ also, and this shows that $m(z; H) = u(0, z) = m_J(z)$. □

So all Jacobi m functions m_J can be realized by a canonical system, and these are exactly the Herglotz functions of the form

$$m_J(z) = \int \frac{d\rho(t)}{t-z},$$

with a compactly but not finitely supported probability measure ρ. We then also obtain all functions Am_J, $A \in SL(2,\mathbb{R})$, from Theorem 3.20, and we can add a term bz, with $b > 0$, by starting with a singular interval of type e_2 of that length. It follows that we can realize any $F \in \mathcal{F}$ of the form

$$F(z) = a + bz + c\int \frac{d\rho(t)}{t-z},$$

with ρ as above and $a \in \mathbb{R}$, $b \geq 0$, $c > 0$. Clearly, these are dense in $\mathcal{F}$. This concludes the description of the second approach, and we could now finish the proof of the existence part of Theorem 5.1 exactly as above.

5.4 The ordering theorem for de Branges spaces

In this section, we will prove an important structural result about de Branges spaces. It will power the proofs of the uniqueness parts of Theorems 5.1, 5.2, and it is of considerable independent interest. It is probably the crowning achievement of de Branges's whole theory.

Theorem 5.11. *Let $B(E_j)$, $j = 1,2$, be regular de Branges spaces which are isometrically contained in $L^2(\mathbb{R},\rho)$, for some Borel measure ρ. Then one of the two de Branges spaces is isometrically contained in the other.*

The assumption that $B(E_j) \subseteq L^2(\mathbb{R},\rho)$ isometrically is of course given its obvious interpretation: for every function $F \in B(E_j)$, its restriction to $\mathbb{R}$ belongs to $L^2(\mathbb{R},\rho)$, and $\|F\|_{B(E_j)} = \|F\|_{L^2(\mathbb{R},\rho)}$.

Recall also that regularity of a de Branges space was defined in terms of the equivalent conditions of Theorem 4.24.

We can in particular take $d\rho(t) = dt/(\pi|E(t)|^2)$, with E being a given de Branges function, and then the theorem says that the de Branges subspaces of a given de Branges space $B(E)$ are totally ordered by inclusion.

As a preparation for the proof, we observe the following properties of ρ.

Lemma 5.12. *Suppose that $L^2(\mathbb{R},\rho)$ isometrically contains the regular de Branges space $B(E)$. Then*

$$\int_{-\infty}^{\infty} \frac{1+|E(t)|^2}{1+t^2}\, d\rho(t) < \infty.$$

Proof. Let us first assume that E has a zero w, which then necessarily lies in $\mathbb{C}^-$. By Theorem 4.24, $S_w E \in B(E)$, and since $|t-w|^2$ and t^2+1 are comparable in size for $t \in \mathbb{R}$, this implies that

$$\int_{-\infty}^{\infty} \frac{|E(t)|^2}{1+t^2}\, d\rho(t) < \infty. \tag{5.7}$$

Similarly, $(S_0 E)(t) = (E(t)-1)/t \in L^2(\rho)$, so

$$\int_{-\infty}^{\infty} \frac{|E(t)-1|^2}{1+t^2}\, d\rho(t) < \infty$$

as well, and by combining this with (5.7), we see that also $1/(1+t^2) \in L^1(\rho)$.

If E is zero-free, $E(0) = 1$, then $E(z) = e^{az}$, with $a \in \mathbb{C}$, from the Hadamard factorization. We have $E \in N$, since $B(E)$ is regular, and of course E is a de Branges function. This further narrows it down to $E(z) = e^{-i\tau z}$, with $\tau > 0$. It is now easy to prove that $1/(1+t^2) \in L^1(\rho)$ in this case also, along the lines of the discussion of the previous paragraph. □

The proof of Theorem 5.11 will be given in several stages, and in the first one, we will establish the asserted inclusion not for the de Branges spaces themselves, but for their images in $L^2(\mathbb{R},\rho)$. For clearer formal organization, let us state this as a separate result.

Proposition 5.13. *In the situation of Theorem 5.11, write $X_j = RB(E_j) \subseteq L^2(\mathbb{R},\rho)$ for the image of $B(E_j)$ under the restriction map $R : B(E_j) \to L^2(\mathbb{R},\rho)$. Then $X_1 \subseteq X_2$ or $X_2 \subseteq X_1$.*

If ρ is not a discrete measure, then any $F \in L^2(\mathbb{R},\rho)$ has at most one entire extension, so in this case, this is the same as Theorem 5.11 itself. In general, however, we will need a separate argument to obtain the theorem from the proposition, and we will in fact follow a somewhat convoluted plan to eventually establish Theorem 5.11 in full generality. We will start out by proving the proposition, and then we analyze its relation to Theorem 5.11 more carefully at the end of this section. Then we use Proposition 5.13 to prove Theorems 5.1, 5.2, and then we can finally return to the present analysis and fully establish Theorem 5.11.

Proof. For $f_1 \in L^2(\mathbb{R},\rho)$, $f_1 \in X_1^\perp$, we define two functions $G_\pm(z)$ by

$$G_+(z) = \int_{-\infty}^{\infty} \frac{E_1(z)E_2(t) - E_1(t)E_2(z)}{t-z} f_1(t)\, d\rho(t),$$

and G_- is defined by the same formula, but with E_1 replaced by $E_1^\#$ throughout. Lemma 5.12 makes sure that these integrals converge for any $z \in \mathbb{C}$ and define entire functions.

Now consider the (meromorphic) function

$$\frac{G_+(z)}{E_1(z)} - \frac{G_-(z)}{E_1^\#(z)}.$$

A straightforward calculation shows that this is a constant multiple of $E_2/(E_1E_1^\#)\langle J_z^{(1)}, f_1\rangle$, and here $J_z^{(1)}$ denotes the reproducing kernel of $B(E_1)$, which is orthogonal to f_1, by the choice of this function. So $G_+/E_1 = G_-/E_1^\#$, and since E_1 and $E_1^\#$ do not have common zeros, this lets us define an entire function G_1 as

$$G_1(z) = \begin{cases} G_+(z)/E_1(z), & \operatorname{Im} z \geq 0, \\ G_-(z)/E_1^\#(z), & \operatorname{Im} z < 0. \end{cases}$$

Similarly, if we let the two de Branges spaces swap roles, we can define a second entire function,

$$G_2(z) = \begin{cases} \dfrac{1}{E_2(z)} \displaystyle\int_{-\infty}^{\infty} \frac{E_2(z)E_1(t) - E_2(t)E_1(z)}{t-z} f_2(t)\, d\rho(t), & \operatorname{Im} z \geq 0, \\ \dfrac{1}{E_2^\#(z)} \displaystyle\int_{-\infty}^{\infty} \frac{E_2^\#(z)E_1(t) - E_2^\#(t)E_1(z)}{t-z} f_2(t)\, d\rho(t), & \operatorname{Im} z < 0, \end{cases}$$

for any $f_2 \in L^2(\mathbb{R}, \rho)$, $f_2 \in X_2^\perp$.

I now claim that G_1, G_2 satisfy the conditions of Krein's Theorem 4.2, that is, $G_j, G_j^\# \in N$. To see this, note that for $z \in \mathbb{C}^+$, the integral may be split into two parts that converge separately, so (for example)

$$G_1(z) = \int_{-\infty}^{\infty} \frac{E_2(t)}{t-z} f_1(t)\, d\rho(t) - \frac{E_2(z)}{E_1(z)} \int_{-\infty}^{\infty} \frac{E_1(t)}{t-z} f_1(t)\, d\rho(t).$$

Since $E_2/E_1 \in N$, by Theorem 4.19, it now suffices to show that the integrals define functions from N also. This is obvious because they are of the form

$$\int_{-\infty}^{\infty} \frac{g(t)}{t-z}\, d\rho(t), \qquad \int_{-\infty}^{\infty} \frac{|g(t)|}{1+|t|}\, d\rho(t) < \infty$$

(by Lemma 5.12 again), so are linear combinations of (at most) four Herglotz functions, by writing g as a combination of four non-negative functions. The proof that $G_j^\# \in N$ is of course completely analogous.

We can extract additional information on the asymptotics here by using the Cauchy–Schwarz inequality, so for example

$$\left| \int_{-\infty}^{\infty} \frac{E_2(t)}{t-z} f_1(t)\, d\rho(t) \right|^2 \leq \|f_1\|^2 \int_{-\infty}^{\infty} \frac{|E_2(t)|^2}{|t-z|^2}\, d\rho(t).$$

For $z = Re^{i\theta}$, $0 < \theta < \pi$, we have

$$|t - z|^2 = t^2 + R^2 - 2tR\cos\theta \geq (1 - |\cos\theta|)(t^2 + R^2),$$

and $1 - |\cos\theta| \geq (1/2)\sin^2\theta$, so we can estimate the integral as follows:

$$\int_{-\infty}^{\infty} \frac{|E_2(t)|^2}{|t-z|^2}\, d\rho(t) \leq \frac{2}{\sin^2\theta} \int_{-\infty}^{\infty} \frac{|E_2(t)|^2}{t^2 + R^2}\, d\rho(t) = \frac{o(1)}{\sin^2\theta}, \quad R \to \infty.$$

By using a similar bound on the other integral and conducting the same analysis in the lower half plane, we obtain the bound

$$|G_1(z)| \leq \begin{cases} \dfrac{o(1)}{\sin\theta}\left(1 + \left|\dfrac{E_2(z)}{E_1(z)}\right|\right), & z = Re^{i\theta} \in \mathbb{C}^+, \\ \dfrac{o(1)}{|\sin\theta|}\left(1 + \left|\dfrac{E_2(z)}{E_1^\#(z)}\right|\right), & z = Re^{i\theta} \in \mathbb{C}^-. \end{cases} \tag{5.8}$$

Of course, there is a similar bound on G_2, which we derive in the same way; it reads

$$|G_2(z)| \leq \begin{cases} \dfrac{o(1)}{\sin\theta}\left(1 + \left|\dfrac{E_1(z)}{E_2(z)}\right|\right), & z = Re^{i\theta} \in \mathbb{C}^+, \\ \dfrac{o(1)}{|\sin\theta|}\left(1 + \left|\dfrac{E_1(z)}{E_2^\#(z)}\right|\right), & z = Re^{i\theta} \in \mathbb{C}^-. \end{cases} \tag{5.9}$$

Let $\tau_j = \tau(E_j) \geq 0$ be the mean types of E_1, E_2. We now distinguish two cases. (1) Assume that $\tau_1 \neq \tau_2$ and let's say $\tau_1 > \tau_2$. Then $|E_2/E_1|$ decays exponentially along any ray $Re^{i\theta}$, $0 < \theta < \pi$, as $R \to \infty$. Thus $G_1(Re^{i\theta}) \to 0$ also for these θ, by (5.8). The same observation applies to the lower half plane $z \in \mathbb{C}^-$ since $|E_2(z)| < |E_2^\#(z)|$ there. So $G_1(Re^{i\theta}) \to 0$ along any non-real ray. We now use the following version of the Phragmen–Lindelöf principle to conclude that $G_1 \equiv 0$.

Theorem 5.14 (Phragmen–Lindelöf). *Suppose that $f : \overline{S} \to \mathbb{C}$ is holomorphic on the open sector $S = \{Re^{i\theta} : R > 0, \alpha < \theta < \beta\}$ and continuous on $\overline{S}$. Suppose further that $\beta - \alpha < \pi$, $|f(z)| \leq e^{\tau|z|}$ for some $\tau > 0$, and $|f(z)| \leq M$ on the boundary of S. Then $|f(z)| \leq M$ for all $z \in S$.*

Proof. We can assume, by a change of variable, that the sector is of the form $-\alpha < \theta < \alpha$, with $0 < \alpha < \pi/2$. Fix a $k > 1$ with $k\alpha < \pi/2$, let $\epsilon > 0$, and consider

$$g(z) = f(z)e^{-\epsilon z^k};$$

here we specifically take the (holomorphic on S) determination $z^k = R^k e^{ik\theta}$ of the kth power. Then $\operatorname{Re} z^k = R^k \cos k\theta \geq \delta R^k$, uniformly on S, for some $\delta > 0$. Thus $g(Re^{i\theta}) \to 0$ as $R \to \infty$, uniformly in θ.

We now apply the maximum principle to g on $S \cap \{|z| \leq R\}$. By what we just observed, this shows that $|g| \leq M$ on this set. Since $\epsilon > 0$ was arbitrary, it then follows that $|f| \leq M$ also, and this holds everywhere on S since R was also arbitrary. □

To apply this result to G_1, we divide the plane into sectors of opening $< \pi$ (for example, into three of these). Recall that $G_1, G_1^\# \in N$, so G_1 is of exponential type, as required in the Phragmen–Lindelöf principle. It follows that G_1 is bounded, hence constant, and since G_1 converges to zero on rays, the constant is zero.

This is true for any choice of $f_1 \in X_1^\perp$ in the definition of G_1. It follows that the complex conjugate of the other factor in the integral lies in X_1, for any choice of $z \in \mathbb{C}$. Moreover, this is true for both auxiliary functions $G_\pm$ that may be used to define G_1. In other words, we have seen that the following functions of $t \in \mathbb{R}$ lie in X_1, for any $z \in \mathbb{C}$:

$$\frac{\overline{E_1(z)}E_2^\#(t) - E_1^\#(t)\overline{E_2(z)}}{t - \overline{z}}, \quad \frac{\overline{E_1^\#(z)}E_2^\#(t) - E_1(t)\overline{E_2(z)}}{t - \overline{z}}. \tag{5.10}$$

In any de Branges space $B(E)$, if $F \in B(E)$, then, trivially, $F^\# \in B(E)$ also. After restriction, this means that X_1 is invariant under taking complex conjugates. If we apply this observation to the second family of functions from (5.10) and also change z to $\overline{z}$, then we see that also

$$\frac{\overline{E_1(z)}E_2(t) - E_1^\#(t)\overline{E_2^\#(z)}}{t - \overline{z}} \in X_1,$$

for any $z \in \mathbb{C}$. By taking a suitable linear combination with the first function from (5.10), it then follows that

$$\overline{E_1(z)}\,\frac{\overline{E_2^\#(z)}E_2^\#(t) - \overline{E_2(z)}E_2(t)}{t - \overline{z}} \in X_1.$$

This is a multiple of $J_z^{(2)}$, so $J_z^{(2)} \in X_1$ whenever $E_1(z) \neq 0$, and these functions span X_2, so $X_2 \subseteq X_1$, as claimed.

(2) It remains to discuss the more difficult case when $\tau_1 = \tau_2$. Now it will be technically convenient to gain some extra decay by introducing the new functions

$$H_j(z) = \frac{G_j(z) - G_j(0)}{z}.$$

These then obey the bounds (writing, as usual, $z = x + iy$)

$$|H_1(z)| \lesssim \begin{cases} \frac{1}{y}\left(1 + \left|\frac{E_2(z)}{E_1(z)}\right|\right), & z \in \mathbb{C}^+, \\ \frac{1}{|y|}\left(1 + \left|\frac{E_2(z)}{E_1^\#(z)}\right|\right), & z \in \mathbb{C}^-, \end{cases} \tag{5.11}$$

$$|H_2(z)| \lesssim \begin{cases} \frac{1}{y}\left(1 + \left|\frac{E_1(z)}{E_2(z)}\right|\right), & z \in \mathbb{C}^+, \\ \frac{1}{|y|}\left(1 + \left|\frac{E_1(z)}{E_2^\#(z)}\right|\right), & z \in \mathbb{C}^-. \end{cases} \tag{5.12}$$

Since $\log|E_j(iR)| = \tau R + o(R)$ as $R \to \infty$, with $\tau = \tau_1 = \tau_2$, it follows that the mean types of H_1, H_2 in the upper half plane are both non-positive. In the lower half plane, we can estimate $|E_j(z)| < |E_j^{\#}(z)|$ in (5.11), (5.12), so the mean types of $H_j^{\#}$ are also non-positive. Theorem 4.3 now shows that both H_1 and H_2 are of exponential type with type zero (and it would also be quite easy to show that the mean types of H_j, $H_j^{\#}$ are all equal to zero, but we will not need this).

This time, we will need to study the asymptotics of both functions simultaneously, and the important thing will be that

$$\min\{|H_1(z)|, |H_2(z)|\} \lesssim \frac{1}{|y|}, \quad z = x + iy, \tag{5.13}$$

as is immediate from (5.11), (5.12), if we again make use of the fact that $|E_j| < |E_j^{\#}|$ on the lower half plane. We will use this information to bound $\int_0^{2\pi} u_j^2(Re^{i\theta})\, d\theta$ from below, with $u_j = \log^+|H_j|$. The elementary but extremely powerful method we are going to employ is due to Carleman, and its adaptation to our present needs is due to de Branges.

The functions u_j are subharmonic and (obviously) non-negative, and to see the essence of the method, it is best to first discuss it for such a function u that is also smooth ($u_j = \log^+|H_j|$ may or may not have this additional property). The key estimate (5.14) below will say that if u is frequently zero but not identically equal to zero, then (paradoxically perhaps, but of course this is just the kind of thing that typically happens with entire functions) $\int u^2(Re^{i\theta})\, d\theta$ must grow rather fast. We will measure the frequency with which u is zero by introducing

$$L(R) = \left|\{\theta \in [0, 2\pi) : u(Re^{i\theta}) > 0\}\right|.$$

The set on the right-hand side is open, so is a union of arcs, and $L(R)$ is its length.

Lemma 5.15. *Let $u : \mathbb{C} \to [0, \infty)$ be subharmonic, $u \in C^\infty$, and write*

$$Q(R) = \left(\int_0^{2\pi} u^2(Re^{i\theta})\, d\theta\right)^{1/2}.$$

If $Q(a) > 0$ for some $a > 0$, then

$$Q^2(R) \ge 2aQ(a)Q'(a) \int_a^R \exp\left(\int_a^r \alpha(t)\, dt\right) \frac{dr}{r} \tag{5.14}$$

for all $R \ge a$, with

$$\alpha(r) = \begin{cases} \dfrac{2\pi}{rL(r)}, & L(r) < 2\pi, \\ 0, & L(r) = 2\pi. \end{cases}$$

Before we prove this, it might be helpful to quickly review those properties of subharmonic functions that will play a role below. These functions have the submean property, that is,

$$u(z) \le \frac{1}{2\pi} \int_0^{2\pi} u(z + Re^{i\theta})\, d\theta,$$

and their Laplacian Δu is non-negative. These two properties are in fact equivalent, and either one could serve as the defining property, certainly for smooth functions, but also for general upper semi-continuous functions, if some care is exercised with the interpretation of the Laplacian. In polar coordinates, the Laplacian is given by

$$(\Delta u)(re^{i\theta}) = u_{rr} + \frac{1}{r} u_r + \frac{1}{r^2} u_{\theta\theta}.$$

The circular means are an increasing function of R, and a convex function of $\log R$. Finally, if $u \ge 0$ is subharmonic, then so is u^2; the general fact behind this remark is that one may apply an increasing convex function.

In particular, these remarks show that $Q(R)$ also is an increasing function, and thus $L(t) > 0$ for $t \ge a$, so everything in (5.14) is well defined.

Proof. We compute $QQ' = \int_0^{2\pi} uu_r\, d\theta$, so

$$(rQ(r)Q'(r))' = \int_0^{2\pi} (uu_r + ru_r^2 + ruu_{rr})\, d\theta.$$

Since $\Delta u \ge 0$, the integrand is $\ge ru_r^2 - (1/r)uu_{\theta\theta}$, and since u is a periodic function of θ, an integration by parts shows that

$$(rQQ')' \ge \int_0^{2\pi} \left(ru_r^2 + \frac{1}{r} u_\theta^2 \right) d\theta. \tag{5.15}$$

If now $L(r) < 2\pi$, then we use Wirtinger's inequality, stated as Lemma 5.16 below, to estimate the second term. As already observed above, the (non-empty, if $r \ge a$) set $\{u > 0\}$ is a union of intervals I_j of lengths L_j, possibly infinitely many in our current setting (though not in the eventual application of the method to $u = \log^+ |H_k|$), with $\sum L_j = L$. We apply Wirtinger's inequality to each of these to obtain

$$\int_{I_j} u_\theta^2\, d\theta \ge \frac{\pi^2}{L_j^2} \int_{I_j} u^2\, d\theta,$$

and since $L_j \le L = L(r)$ for all j, summation over j then yields

$$\int_0^{2\pi} u_\theta^2\, d\theta \ge \frac{\pi^2}{L^2} Q^2.$$

We also estimate the first term from the right-hand side of (5.15) by observing that

$$(QQ')^2 = \left(\int_0^{2\pi} uu_r \, d\theta \right)^2 \leq Q^2 \int_0^{2\pi} u_r^2 \, d\theta,$$

so $\int u_r^2 \geq Q'^2$, at least if $Q(r) > 0$, but we already observed that this will hold for $r \geq a$. By plugging these bounds into (5.15), we see that

$$(rQQ')' \geq rQ'^2 + \frac{\pi^2 Q^2}{rL(r)^2} \geq \frac{2\pi QQ'}{L(r)}.$$

In other words,

$$(rQQ')' \geq \alpha(r) rQQ', \tag{5.16}$$

and this final estimate is also trivially true in the other case, when $L(r) = 2\pi$.

Recall also that Q is an increasing function of r, so $rQQ' \geq 0$, and thus (5.16) implies that

$$rQ(r)Q'(r) \geq aQ(a)Q'(a) \exp\left(\int_a^r \alpha(t) \, dt \right).$$

To derive this, it is tempting to just divide through by rQQ', which makes the left-hand side of (5.16) equal to $(\log rQQ')'$, and integrate, but of course this runs into the issue of possible zeros of rQQ'. It is thus cleaner from a formal point of view to refer to a version of Gronwall's lemma to reach the desired conclusion; the one needed here differs from the one discussed in Chapter 4 in that we now have a *lower* bound on our function, but that does not introduce any problems.

Finally, we divide through by r and integrate to obtain the asserted bound (5.14); an additional constant $Q^2(a) > 0$ on the right-hand side has been dropped. □

Lemma 5.16 (Wirtinger). *Suppose that $f \in C^1[a,b]$ with $f(a) = f(b) = 0$. Then*

$$\int_a^b |f(x)|^2 \, dx \leq \left(\frac{b-a}{\pi} \right)^2 \int_a^b |f'(x)|^2 \, dx.$$

Proof. It suffices to prove this for $a = 0$, $b = \pi$. Extend f to $[-\pi, \pi]$ by making it an odd function; this is still a C^1 function, on the circle. Of course, the original inequality, with constant 1, is equivalent to the inequality for the extended function, still with constant 1. By Parseval's identity, this is equivalent to the corresponding inequality

$$\sum_{n=-\infty}^{\infty} |\widehat{f}_n|^2 \leq \sum_{n=-\infty}^{\infty} n^2 |\widehat{f}_n|^2$$

on the Fourier coefficients of f. Since f is odd now, we have $\widehat{f}_0 = 0$, so this version is obvious. □

Now we adapt Lemma 5.15 to what will be our actual application of the method.

Lemma 5.17. *Let F be a non-constant entire function. Then, with the same definitions as above, the subharmonic function $u = \log^+ |F|$ also satisfies*

$$Q^2(R) \geq c \int_1^R \exp\left(\int_1^r \alpha(t)\, dt \right) \frac{dr}{r}, \tag{5.17}$$

for some $c > 0$ and for all large $R > 1$.

Proof. Let us first assess how far we are from the situation of Lemma 5.15. Of course, u is smooth off the (closed) set $\{z : |F(z)| = 1\}$. Moreover, $u^2 \in C^1$, as is easy to confirm by using the Taylor expansion of F about points z_0 with $|F(z_0)| = 1$. If we do not take the square, then what we can say is that u satisfies a Lipschitz condition on every compact set.

Finally, we may and will assume that $\{z : |z| = r, |F(z)| = 1\}$ is finite for every $r \geq 0$. To see this, consider the function

$$g(z) = F(z)\overline{F(r^2/\overline{z})}.$$

Since g is holomorphic on $\mathbb{C} \setminus \{0\}$ and $g(z) = |F(z)|^2$ for $|z| = r$, the presence of infinitely many points with $|F| = 1$ on this circle will make $g \equiv 1$, and then $|F(z)| \lesssim |z|^N$ as $|z| \to \infty$, so F is a polynomial. In this case, however, (5.17) is easily established directly: if F is a non-constant polynomial, then $\alpha(t) = 0$ for all large t, so the right-hand side of (5.17) is $\simeq \log R$, while $Q^2(R) \simeq \log^2 R$, so (5.17) holds, with room to spare.

We now define smooth approximations of u in the usual way by taking convolutions with a smooth approximation to the identity, and with additional attention to detail exercised here, we can also take advantage of the fact that u is subharmonic. More specifically, we set

$$u_n(z) = \int_0^\infty dr\, \varphi(r) \int_0^{2\pi} \frac{d\theta}{2\pi}\, u\left(z + \frac{r}{n} e^{i\theta}\right),$$

and here φ is a smooth compactly supported function with $\int \varphi = 1$, $\varphi \geq 0$. Clearly, $u_n \in C^\infty$ and $u_n \to u$ locally uniformly. Moreover, since each u_n is an average of subharmonic functions, it inherits the submean property from those and thus is subharmonic itself. More precisely, u_n is an average (with respect to the radial variable) of circular averages of u, and these are $\geq u(z)$, so we also see that $u_n \geq u$.

Now we of course apply Lemma 5.15 to u_n, so have (5.14) available for these functions, and then send $n \to \infty$. On the left-hand side, the convergence $Q_n^2(R) \to Q^2(R)$ is obvious. On the right-hand side, we need to address two issues: we must bound the constants $(Q_n^2)'(a)$ from below, and we must relate the integrals with α_n to those with α.

To control $(Q_n^2)'(a) = 2\int_0^{2\pi} u_n (u_n)_r \, d\theta$, we first recall that $u^2 \in C^1$, so $Q^2 \in C^1$ also and

$$(Q^2)'(a) = 2\int_0^{2\pi} (u^2)_r \, d\theta = 2\int_0^{2\pi} u u_r \, d\theta;$$

the last step makes use of our assumption that there are only finitely many points with $|F| = 1$ on our circle, at which u might fail to be differentiable. Now dominated convergence shows that $(Q_n^2)'(a) \to (Q^2)'(a)$. So we only need to find an a with $(Q^2)'(a) > 0$. If there is no such a, then $Q^2(r)$ is constant. For any subharmonic function, we can average the submean property over the radial variable also, and thus $u(z)$ is also bounded by the average over any disk $\{w : |w - z| \le r\}$. Thus if Q^2 is bounded, then so is u, but this makes $|F|$ bounded and thus constant.

Now let us look at what happens to the integrals from (5.14) when we send $n \to \infty$. By two applications of Fatou's lemma, it actually suffices to show that

$$\liminf_{n\to\infty} \alpha_n(t) \ge \alpha(t)$$

for almost every $t > 0$ (but we will show it for all $t > 0$). Or, if we say this in terms of $L_n(t)$ and $L(t)$, then what we want to show becomes

$$\liminf_{n\to\infty} \left|\{\theta \in [0, 2\pi) : u_n(te^{i\theta}) = 0\}\right| \ge \left|\{\theta \in [0, 2\pi) : u(te^{i\theta}) = 0\}\right|.$$

This is fairly obvious: the set $\{u = 0\}$ consists of the finitely many points where $|F| = 1$, plus possibly some arcs connecting those, and $|F| < 1$ on these arcs. So removal of a small neighborhood of the points where $|F| = 1$ will give us a set $S \subseteq \{z : |z| = t\}$ of almost full measure in $\{u = 0\}$, but now $\sup_S |F| < 1$. So we also have $|F| < 1$ on an open (in $\mathbb{C}$) neighborhood of this set, and thus $u = 0$ there. Thus, by the definition of the u_n as averages of u over small sets, we also obtain $u_n = 0$ on S for all large n. □

We now have all the tools available to finish the proof of Proposition 5.13. Let

$$Q_j^2(R) = \int_0^{2\pi} \left(\log^+ |H_j(Re^{i\theta})|\right)^2 d\theta.$$

By Lemma 5.17, if neither H_1 nor H_2 is constant, then

$$Q_1^2(R) + Q_2^2(R) \ge c \int_1^R \left(e^{\int_1^r \alpha_1(t)\, dt} + e^{\int_1^r \alpha_2(t)\, dt}\right) \frac{dr}{r}$$

for all large R. We now further estimate the integrand. By the convexity of the exponential function, we have

$$e^{\int_1^r \alpha_1(t)\, dt} + e^{\int_1^r \alpha_2(t)\, dt} \ge 2\exp\left(\int_1^r \frac{\alpha_1(t) + \alpha_2(t)}{2}\, dt\right),$$

and now we take a closer look at $\alpha_1+\alpha_2$. The key estimate (5.13) tells us that if we avoid two small neighborhoods of $\theta=0$ and $\theta=\pi$ of total length C/t, with $C>0$ suitably chosen, then at least one of H_1 or H_2 will be less than one in absolute value on the remaining part of the circle of radius t. In other words,

$$L_1(t)+L_2(t)\le 2\pi+\frac{C}{t}$$

for all large t. Let us first deal with the case when $L_1(t), L_2(t)<2\pi$. Then

$$\alpha_1(t)+\alpha_2(t)=\frac{2\pi}{t}\left(\frac{1}{L_1(t)}+\frac{1}{L_2(t)}\right),$$

and by using the elementary inequality $1/a+1/b\ge 4/(a+b)$ here, this produces

$$\alpha_1(t)+\alpha_2(t)\ge\frac{8\pi}{t(L_1(t)+L_2(t))}\ge\frac{4}{t}-O(1/t^2). \tag{5.18}$$

This final bound is also valid in the other case, when $L_j=2\pi$ for some j, and let us say $L_1=2\pi$, so $\alpha_1=0$: in this situation, (5.13) shows that $|H_2|\le 1$ on almost the whole circle, again with the possible exception of two small arcs about $\theta=0$, $\theta=\pi$ of total length $\lesssim 1/t$. So $L_2(t)\lesssim 1/t$, and thus $\alpha_2\gtrsim 1$, which is actually quite a bit more than what I asserted in (5.18).

So, putting everything together, we now see that

$$\begin{aligned}Q_1^2(R)+Q_2^2(R)&\ge 2c\int_1^R\exp\left(\int_1^r\left(\frac{2}{t}-\frac{C}{t^2}\right)dt\right)\frac{dr}{r}\\&\ge c_1\int_1^R r^2\,\frac{dr}{r}\gtrsim R^2,\end{aligned}$$

but this contradicts our earlier result that H_1, H_2 have exponential type zero.

The only way to avoid this contradiction is to admit that at least one of these functions is constant, and let us say that $H_1\equiv c$. Then G_1 is a polynomial (of degree at most 1, but we do not need this extra information here). Now we return to (5.8). If G_1 is not the zero polynomial, then $|G_1(z)|\gtrsim 1$ for all large z, and this forces $|E_2/E_1|$ to become large along all rays $Re^{i\theta}$, $0<\theta<\pi$, in the upper half plane, and similarly $|E_2/E_1^\#|\to\infty$ along rays in the lower half plane. Now (5.9) shows that then $G_2\to 0$ along all these rays, so $G_2\equiv 0$, by the Phragmen–Lindelöf principle.

To summarize: we have shown that for any choice of $f_1\in X_1^\perp$, $f_2\in X_2^\perp$ as above, at least one of the two functions G_1, G_2 is identically equal to zero. Since we can vary f_1, f_2 independently here, it must then actually be the case that either $G_1\equiv 0$ for all f_1, or $G_2\equiv 0$ for all f_2. This then finally allows us to finish the proof as in case (1) above. □

Now let us try to bridge the gap between the proposition and Theorem 5.11. So suppose the situation is as described in the statement of Proposition 5.13: we have two subspaces $X_1 \subseteq X_2 \subseteq L^2(\mathbb{R},\rho)$, and these arose as restrictions of functions from two regular de Branges spaces $B(E_j)$, $j = 1,2$. We wish to identify conditions which will then guarantee that $B(E_1) \subseteq B(E_2)$ also, as entire functions.

Since $S_wE_1 \in B(E_1)$, so the restriction to $\mathbb{R}$ lies in X_1, we see that $(S_wE_1)(z)$ agrees ρ-almost everywhere with a function $F_w \in B(E_2)$, and this is true for all $w \in \mathbb{C}$. Let us focus specifically on $w = 0$ for the moment. Then we can define an entire function P as $P(z) = zF_0(z) + 1$, and this is done so that then $S_0P \in B(E_2)$ and $P = E_1$, $S_0P = S_0E_1$ almost everywhere with respect to ρ.

I now claim that then also $S_wP \in B(E_2)$ for arbitrary $w \in \mathbb{C}$, and this follows from an argument that we have already seen a number of times, in the proofs of Theorem 4.4 and especially Theorem 4.24, but let me sketch it anyway, as a quick reminder: we must check that $(S_wP)/E_2 \in H^2$ and $(S_wP)^\#/E_2 \in H^2$. When we verify these properties, using the defining condition (4.1) of Chapter 4 of H^2, it is easy to confirm that there are no problems for $z = t + iy$ near w, and for $|z - w| \geq \delta$, we can write

$$\frac{(S_wP)(z)}{E_2(z)} = \frac{P(z) - P(0)}{(z - w)E_2(z)} + \frac{P(0) - P(w)}{(z - w)E_2(z)},$$

and now the first term on the right-hand side satisfies an H^2 type condition (but with z restricted to $|z - w| \geq \delta$) because $(S_0P)/E_2 \in H^2$, and for the second one, we use the fact that $1/[(z + i)E_2] \in H^2$.

Now we repeat this whole process of going from E_1 to P, taking the new function $E_1 - P$ as our starting point, and we also do this for arbitrary $w \in \mathbb{C}$ now. So consider the function $S_w(E_1 - P)$. It lies in X_2 after restricting, being the difference of two such functions, so agrees ρ-almost everywhere with a function $G_w \in B(E_2)$, and we can again define a new entire function

$$Q_w(z) = (z - w)G_w(z) + E_1(w) - P(w). \tag{5.19}$$

This has the same general properties as P above: almost everywhere with respect to ρ, we have $Q_w = E_1 - P = 0$, $S_wQ_w = S_w(E_1 - P)$, and $S_wQ_w \in B(E_2)$. Moreover, in exactly the same way as above, we can then show that also $S_vQ_w \in B(E_2)$ for arbitrary $v \in \mathbb{C}$.

As we will see in a moment, the existence of a non-trivial entire function Q with these two properties,

$$Q(t) = 0 \text{ for } \rho\text{-a.e. } t \in \mathbb{R}, \quad S_vQ \in B(E_2) \text{ for all } v \in \mathbb{C}, \tag{5.20}$$

will put us in a rather special situation. So before we study these, let us see what we would get from the additional assumption (on $B(E_2)$ and ρ) that there is no entire function $Q \not\equiv 0$ satisfying (5.20). In this case, (5.19) for $z = w$ implies that $E_1(w) = P(w)$. Here we can take any $w \in \mathbb{C}$, so $E_1(z) = P(z)$, and thus $(S_wE_1)(z) \in B(E_2)$, without the need for any modification off the support of ρ. We can then obtain the reproducing kernels of $B(E_1)$ as linear combinations of these functions S_wE_1, $(S_wE_1)^\#$, so these will be in $B(E_2)$, too, and thus $B(E_1) \subseteq B(E_2)$ in this case, as desired.

Now let us look at the other case, when there is a non-constant Q satisfying (5.20). Denote the discrete support of ρ by T and consider S_tQ specifically for $t \in T$. Clearly, $(S_tQ)(s) = 0$ if $s \in T$ also, $s \neq t$; at t itself, S_tQ equals $Q'(t)$. This must be non-zero, or else $S_tQ = 0$ almost everywhere, but since the restriction map from $B(E_2)$ to $L^2(\mathbb{R}, \rho)$ is isometric, this means that $(S_tQ)(z) \equiv 0$, so $Q \equiv 0$ after all. So $Q'(t) \neq 0$, and thus S_tQ restricts to a non-zero multiple of δ_t. These functions span $L^2(\mathbb{R}, \rho)$, so $X_2 = L^2(\mathbb{R}, \rho)$.

Here we stop, for now. We are rather close to a full proof of Theorem 5.11, and we will bridge the remaining gap in Section 5.6, after applying what we already have to the inverse spectral problem for canonical systems. For future reference, let us summarize what the situation is at this point.

Proposition 5.18. *Let $X_1 \subseteq X_2 \subseteq L^2(\mathbb{R}, \rho)$ be subspaces of an $L^2(\mathbb{R}, \rho)$ space, which were obtained by isometrically restricting the functions from two regular de Branges spaces $B(E_j)$, $j = 1, 2$. If $X_2 \neq L^2(\mathbb{R}, \rho)$, then also $B(E_1) \subseteq B(E_2)$.*

This is what we showed above. The analysis of the other case, when $X_2 = L^2(\mathbb{R}, \rho)$, could be continued, along the lines of what we did in the proof of Theorem 4.10, and one can show that the function Q from above must be a multiple of $A_2(z)\sin\beta - C_2(z)\cos\beta$, for some β, and with $E_2 = A_2 - iC_2$. It is not necessary for us here to do this at this point because we will soon connect arbitrary regular de Branges spaces to canonical systems, and then we can refer to Theorem 4.10 directly (rather than repeat the underlying argument in the present more abstract scenario).

5.5 Existence of a canonical system with given transfer matrix

We prove the existence part of Theorem 5.2 here. Let $T \in TM$ be given, and denote its matrix elements by A, B, C, D, as usual. We can in fact assume that $T \in TM_1$ because otherwise Theorem 4.17 delivers an $H \in \mathcal{C}(L)$ (consisting of a single singular interval) with $T(z) = T(L; z; H)$ for free. Then $E = A - iC$ is a de Branges function, the de Branges space $B(E)$ is regular, so in particular

$$\int_{-\infty}^{\infty} \frac{dt}{(1+t^2)|E(t)|^2} < \infty,$$

by Theorems 4.18 and 4.19. We can thus define a Herglotz function M as

$$M(z) = \int_{-\infty}^{\infty} \left(\frac{1}{t-z} - \frac{t}{t^2+1} \right) \frac{dt}{\pi |E(t)|^2}.$$

By the already established existence part of Theorem 5.1, there is an $H \in \mathcal{C}$ with $m(z;H) = M(z)$. Its spectral measure is given by

$$d\rho(t) = \frac{dt}{\pi|E(t)|^2}.$$

In particular, $B(E)$ is isometrically contained in $L^2(\mathbb{R},\rho)$. Moreover, the canonical system H delivers a chain of regular de Branges spaces $B(E_x)$, $E_x(z) = u_1(x,z) - iu_2(x,z)$, one space for each regular point $x \in (0,\infty)$, as in Theorem 4.8(b), and these are also isometrically contained in $L^2(\mathbb{R},\rho)$.

We will now apply the ordering theorem, in the form of Proposition 5.13, to this situation, but before we do this, it might be helpful to quickly review the reasoning behind Theorem 4.8, since these arguments will play a central role in what follows.

Let us denote the set of regular points of the canonical system $H \in \mathcal{C}$ by $R \subseteq (0,\infty)$. Then, for $L \in R$, we introduced the spaces $\mathcal{H}(L) = L^2_H(0,L) \ominus \mathcal{S}_0(0)$, and here $\mathcal{S}_0$ is the restriction of the maximal relation on $(0,L)$ defined by the conditions $u_2(0) = u(L) = 0$. The space $\mathcal{H}(L)$ has an easy direct description as

$$\mathcal{H}(L) = L^2_H(R \cap (0,L)) \oplus \bigoplus_j L(\chi_{(c_j,d_j)} e_{\alpha_j});$$

the sum is over the singular intervals $(c_j,d_j) \subseteq (0,L)$, and the types of these have been denoted by α_j. See also Theorem 3.19. (There would be a small modification to this formula if $(0,L)$ started with a singular interval of type e_2, but that would make $b(M) > 0$, by Theorem 4.33, so this will not happen here; similarly, a singular half line would be inconsistent with the non-discrete spectral measure.)

It is now clear that $\mathcal{H}(L_1) \subseteq \mathcal{H}(L_2)$ isometrically if $L_1, L_2 \in R$, $L_1 \le L_2$, and since

$$(Uf)(z) = \int_0^\infty u^*(x,\overline{z})H(x)f(x)\,dx$$

maps $\mathcal{H}(L)$ unitarily onto $B(E_L)$ (Theorem 4.7) and isometrically into $L^2(\mathbb{R},\rho)$ (Theorem 3.17), we obtain the asserted isometric inclusions for the de Branges spaces $B(E_{L_j})$.

Now let us return to the business at hand. Given $T \in TM_1$, we have constructed a canonical system $H \in \mathcal{C}$ with spectral measure $d\rho = dt/(\pi|E|^2)$, and for any $x \in R$, we know that both $B(E_x)$ and $B(E)$ are isometrically contained in $L^2(\mathbb{R},\rho)$. There is one small technical point that perhaps deserves a brief comment before we proceed: E_x indeed is a de Branges function for every $x \in R$, and $B(E_x) \subseteq L^2(\mathbb{R},\rho)$. This is not entirely automatic because $(0,x)$ could be a single singular interval. However, then this singular interval will not be of type e_2, as we just observed, since we took the precaution to make $b(M) = 0$. Thus we are still covered in this case also by Theorems 4.8(c), 4.17, and 4.18.

The measure ρ is absolutely continuous, so there is no difference between Proposition 5.13 and Theorem 5.11 in the current situation. It follows that either $B(E) \subseteq B(E_x)$

or $B(E_x) \subseteq B(E)$, and the inclusion is isometric. Define

$$L_- = \sup\{x \in R : B(E_x) \subseteq B(E)\},$$
$$L_+ = \inf\{x \in R : B(E_x) \supseteq B(E)\};$$

if the first set is empty, then we can set $L_- = 0$, and, similarly, $L_+ = \infty$ if the second one is empty. But actually these scenarios do not occur: if we had $L_- = 0$, then $B(E) \subseteq B(E_x)$ for all $x \in R$, and thus also

$$B(E) \subseteq \bigcap_{x \in R} B(E_x).$$

This intersection is the zero space if there are $x \in R$ arbitrarily close to 0 because then $\bigcap_{x\in R} \mathcal{H}(x) = 0$, but a de Branges space never is the zero space, so this is impossible. So $(0, \infty)$ would have to start with a singular interval $(0, d)$. In this case, the intersection equals $B(E_d)$, as we again see by considering the spaces $\mathcal{H}(x)$ and mapping by U, and this space is one-dimensional, being the unitary image of $L(\chi_{(0,d)} e_\alpha)$ under U. So $B(E) = B(E_d)$, and $d \in R$, so the set of $x \in R$ with $B(E_x) \subseteq B(E)$ was not empty after all, and $L_- = d$ in this case.

In the same way, we can show that $L_+ < \infty$; here, it is important that $B(E) \neq L^2(\mathbb{R}, \rho)$.

So, since the spaces $B(E_x)$ increase as x varies over R, the situation is as follows:

$$0 < L_- \leq L_+ < \infty.$$

We can say more here. First of all, $L_\pm \in R$ since R is closed, and then we also have $B(E_{L_-}) \subseteq B(E)$ and $B(E_{L_+}) \supseteq B(E)$. Let me prove the first of these two claims in detail; the second one will follow from similar arguments. If L_- is in fact a maximum, then this is of course clear. If not, then there are $x \in R$, $x < L_-$, arbitrarily close to L_-, and this implies that

$$B(E_{L_-}) = \overline{\bigcup_{x<L_-, x\in R} B(E_x)},$$

by again noting the analogous property of the $\mathcal{H}(x)$ spaces (which is obvious) and then moving things over to the de Branges spaces with the help of the map U. So $B(E_{L_-}) \subseteq B(E)$ as well, as claimed.

Next, observe that there cannot be any regular points between L_- and L_+, so if $L_- < L_+$, then (L_-, L_+) is a singular interval. This means that $B(E_{L_+}) \ominus B(E_{L_-})$ is one-dimensional, and since

$$B(E_{L_-}) \subseteq B(E) \subseteq B(E_{L_+}),$$

$B(E)$ must be equal to one space or the other. This conclusion is also (trivially) valid if $L_+ = L_-$. So, in all cases, we have found an $L \in R$ such that $B(E) = B(E_L)$.

Now we apply Theorem 4.5: the de Branges functions E and E_L differ by a transformation by a matrix $M \in SL(2,\mathbb{R})$. Since $E(0) = E_L(0) = 1$, this M must have first column e_1. Or, to slightly rephrase what we just derived, for some $a \in \mathbb{R}$, the matrix function

$$T_1(L;z) \equiv \begin{pmatrix} 1 & a \\ 0 & 1 \end{pmatrix} T(L;z;H) \begin{pmatrix} 1 & -a \\ 0 & 1 \end{pmatrix}$$

has the same first column as the $T \in TM_1$ that we were given. By Theorem 1.6, T_1 is the transfer matrix of the modified coefficient function

$$H_1(x) = \begin{pmatrix} 1 & 0 \\ -a & 1 \end{pmatrix} H(x) \begin{pmatrix} 1 & -a \\ 0 & 1 \end{pmatrix}$$

(which may not be trace normed, but we can always fix this at the end and need not worry about it right now). So we have now constructed a canonical system $H_1(x)$ such that its transfer matrix $T(L;z;H_1)$ has the same first column as the given $T \in TM_1$, for suitable $L > 0$. By Theorem 4.22(b), this means that

$$T(z) = T(L;z;H_1) \begin{pmatrix} 1 & bz \\ 0 & 1 \end{pmatrix} \tag{5.21}$$

for some $b \in \mathbb{R}$. If $b \leq 0$ here, then we are done: the matrix product on the right-hand side of (5.21) is the transfer matrix of a singular interval of type e_2 of length $-b$, followed by a shifted version of H_1, and we can then finally run the already announced transformation of the independent variable to make this coefficient function trace normed.

If $b > 0$, then (5.21) says that $a(T(L;z;H_1)) \geq b$, and now we learn from Theorem 4.32 that the trace normed version of H_1 starts with a singular interval of type e_2 of length at least b. So if we write $T(L) = T(L,b)T(b)$, then there will be a cancellation on the right-hand side of (5.21), and in this case, too, $T(z)$ has been realized as the transfer matrix $T_2(L_2;z)$ of a canonical system, with coefficient function $H_2(x) = H_1(x+b)$, $x \geq 0$, and $L_2 = L_1 - b$.

5.6 Uniqueness

We now prove the uniqueness claims of Theorems 5.1, 5.2. Let us start with Theorem 5.1. So suppose that $H_1, H_2 \in \mathcal{C}$ are two trace normed canonical systems with $m(z;H_1) = m(z;H_2)$. If $H \in \mathcal{C}$ is not just a single singular half line, then $m(z;H)$ is a genuine Herglotz function. This means that uniqueness is already clear if $m(z;H) \equiv a \in \mathbb{R}_\infty$: then $H = P_\alpha$, with $a = -\tan\alpha$, as we discussed in the first section of this chapter.

So it suffices to treat the case when the common m function $m(z)$ of H_1, H_2 is a genuine Herglotz function. We can in fact also assume that $b(m) = 0$: if $b(m) > 0$, then Theorem 4.33 shows that both coefficient functions start with a singular interval of type e_2 of this length $b(m)$, and $m(z) = bz + m_1(z)$; here $m_1 \in \mathcal{F}$ denotes the common m function of the shifted coefficient functions $H_j(x+b)$, $x > 0$, and $b(m_1) = 0$ now. So we will now assume that $b(m) = 0$, and thus neither canonical system starts with a singular interval of type e_2.

Denote the common spectral measure by ρ. The key ingredient to the uniqueness proof will be the method of the previous section that allows us to fit any regular de Branges space sitting isometrically inside $L^2(\mathbb{R}, \rho)$ into the chain of de Branges spaces that is delivered by the canonical system. We will now use this to match the de Branges spaces of the two canonical systems. It will be convenient to use the simplified notation $B_j(x) = B(E_x^{(j)})$, $j = 1, 2$, $x \in R_j$, for these spaces.

Let us first assume that at least one of the two systems, let's say H_1, does not end with a singular half line, so has arbitrarily large regular points. Then $B_1(x) \neq L^2(\mathbb{R}, \rho)$ for all $x \in R_1$ because for any such x, there is always an $x' > x$, $x' \in R_1$, that gives us a still larger space. Let $L_2 = \sup R_2 \in (0, \infty]$. Then also $B_2(x) \neq L^2(\mathbb{R}, \rho)$ if $x \in R_2$, $x < L_2$, for the same reason. Fix such an x. By Propositions 5.13, 5.18, for any $t \in R_1$, we then have $B_1(t) \subseteq B_2(x)$ or $B_2(x) \subseteq B_1(t)$ isometrically. We now repeat the arguments of the previous section to find a (unique) $t = t(x) \in R_1$ such that $B_1(t(x)) \equiv B_2(x)$; essentially, $t(x)$ is the t at which the $B_1(t)$ switch from being contained in $B_2(x)$ to containing $B_2(x)$.

Still arguing in the same way as above, we then conclude that there are $a = a(x), b = b(x) \in \mathbb{R}$ such that

$$T_2(x; z) = \begin{pmatrix} 1 & a(x) \\ 0 & 1 \end{pmatrix} T_1(t(x); z) \begin{pmatrix} 1 & -a(x) \\ 0 & 1 \end{pmatrix} \begin{pmatrix} 1 & b(x)z \\ 0 & 1 \end{pmatrix}. \tag{5.22}$$

I now claim that we must have $a = b = 0$ here. To prove this, consider the Weyl disks $\mathcal{D}_j = T_j^{-1}\overline{\mathbb{C}^+}$. What (5.22) says about these is that

$$\mathcal{D}_2(x; z) = \mathcal{D}_1(t(x); z) + a(x) - b(x)z.$$

Recall also that both are genuine disks (not half planes) because neither canonical system starts with a singular interval of type e_2. The common m function $m(z)$ lies in all Weyl disks. If we take specifically $z = iy$ and send $y \to \infty$, then these disks become arbitrarily small, by Theorem 4.30 (and here we again use the fact that there is no initial singular interval of type e_2, to make sure that the assumption of this result is satisfied). Thus they can intersect for all $z = iy$ only if $a = b = 0$, as claimed.

So now we have $T_2(x; z) = T_1(t(x); z)$ for all $x \in R_2$, $x < L_2$. Since $L = -\operatorname{tr} J(\partial T/\partial z)(L; 0)$ for a trace normed canonical system, it follows that $t(x) = x$. So $T_2(x; z) = T_1(x; z)$ for all $x \in R_2$, $x < L_2$. This shows, first of all, that $L_2 = \infty$, or otherwise, since $B_2(L_2) = L^2(\mathbb{R}, \rho)$, we would obtain a space $B_1(x)$ that is either equal to $L^2(\mathbb{R}, \rho)$ after restricting functions (if there are $x \in R_2$, $x < L_2$ arbitrarily close to L_2) or differs from this space by a one-dimensional space (if $(0, L_2)$ ends with a singular interval of H_2). Both scenarios are clearly impossible.

Since $t(x) = x$ was chosen to lie in R_1, we now also see that $R_2 \subseteq R_1$, and then $R_1 = R_2$ since $L_2 = \infty$ means that H_2 can take over the role of H_1 in this whole argument. So both systems have the same singular intervals, and then the types must also agree because for any given singular interval, we already know that the transfer matrices agree at the endpoints. Thus $T_1(x;z) = T_2(x;z)$ for all $x > 0$, and hence $H_1(x) = H_2(x)$ almost everywhere.

It remains to discuss the case when both canonical systems end with a singular half line (L_j, ∞). We can then still match de Branges spaces and proceed as above for all regular points smaller than these L_j. So if (let us say) H_1 has regular points $x \in R_1$, $x < L_1$, but arbitrarily close to L_1, then it follows as above that $L_2 = L_1 =: L$, $H_2 = H_1$ on $(0, L)$. The singular half line (L, ∞) implements a boundary condition, and of course distinct boundary conditions will not lead to identical m functions, so $H_2 = H_1$ on (L, ∞) also.

This finally leaves us with the case where both $(0, L_1)$ and $(0, L_2)$ end with a singular interval (c_j, L_j). Now our method shows that $c_1 = c_2 = c$ and $H_1 = H_2$ on $(0, c)$. For any canonical system, we have $m(z) = T^{-1}(c;z)m_c(z)$, with m_c denoting the m function of the problem on (c, ∞) (or, equivalently, of the shifted coefficient function $H_c(x) = H(x+c), x \geq 0$). So what we have to deal with in this last part is the uniqueness statement of Theorem 5.1, but only for the very special case of systems consisting of exactly two singular intervals. So we now only need to consider $H \in \mathcal{C}$ of the form

$$H(x) = \begin{cases} P_\alpha, & 0 < x < L, \\ P_{\beta^\perp}, & x > L. \end{cases}$$

Its m function is given by

$$m(z; H) = (1 - zLJP_\alpha)e_\beta,$$

and it is easy to show (and even easier to believe) that all three parameters α, β, L can be recovered from m. This completes the proof of Theorem 5.1.

The uniqueness statement of Theorem 5.2 is essentially a special case of this. If $H_1, H_2 \in \mathcal{C}(L)$ share the same transfer matrix $T(L;z)$, then we can extend H_1, H_2 to $(0, \infty)$ by giving them both the same singular half line $H_j = P_\alpha$ on (L, ∞). Then both half line problems have the same m function $m(z) = T^{-1}(L;z)e_{\alpha^\perp}$, so $H_1 = H_2$.

It is now also clear to what extent the spectral measure ρ determines H, and it might be useful to state this separately.

Theorem 5.19. *Two canonical systems from $\mathcal{C}$ have the same spectral measure $\rho \neq 0$ if and only if one coefficient function can be obtained from the other by the following sequence of steps:*

(1) *add or prolong an initial singular interval of type e_2, that is, go from $H(x)$ to*

$$H_1(x) = \begin{cases} P_{e_2}, & 0 < x < b, \\ H(x - b), & x > b; \end{cases}$$

(2) *go from* $H(x)$ *to*

$$H_2(x) = \begin{pmatrix} 1 & 0 \\ a & 1 \end{pmatrix} H(x) \begin{pmatrix} 1 & a \\ 0 & 1 \end{pmatrix},$$

and then to the trace normed version H_3 *of this.*

Similarly, the canonical systems with $\rho = 0$ *are exactly the singular half lines, possibly preceded by a singular interval of type* e_2.

Proof. Obviously, two Herglotz functions m_j have the same ρ if and only if $m_1(z) = m_2(z) + a + bz$ for some $a, b \in \mathbb{R}$, and steps (1), (2) above implement these transformations $m \mapsto m + bz$ and $m \mapsto m + a$, respectively. □

In the course of the uniqueness proof, we essentially established the following result about the de Branges subspaces of $L^2(\mathbb{R}, \rho)$, which can be used to finally finish the proof of Theorem 5.11 and is of great independent interest also.

Theorem 5.20. *Let* $H \in \mathcal{C}$, *with spectral measure* ρ. *If* $B(E)$ *is a regular de Branges space that is isometrically contained in* $L^2(\mathbb{R}, \rho)$, *then* $B(E) \equiv B(E_L)$ *for some regular* $L > 0$.

The converse is also true and was established much earlier, as Theorem 4.8(b), (c), with the (obvious) small caveat that if $(0, d)$ is an initial singular interval of type e_2, then $L = d$ is inadmissible in this converse.

Proof. As I said, the previous discussion has already shown most of this, but let me sketch one more time what we do here: if $B(E) \neq L^2(\mathbb{R}, \rho)$, then Propositions 5.13, 5.18 show that we have to have an isometric inclusion of the spaces $B(E)$, $B(E_x)$, for any regular x with $B(E_x) \neq L^2(\mathbb{R}, \rho)$, one way or the other. The x at which the behavior switches gives us the desired L.

The only situation that needs additional investigation occurs when $B(E) \equiv L^2(\mathbb{R}, \rho)$. We then realize first $E = A - iC$ as the first column of a $T \in TM$ and then this T as the transfer matrix of a second canonical system $H_1 \in \mathcal{C}(L)$, using Theorem 5.2. Now Theorem 4.10 shows that $\rho = \rho_L^{(\beta)}$ must be a spectral measure of this system, for some boundary condition β at $x = L$. Or we can give H_1 the singular half line (L, ∞) of type $\beta^\perp$, and then we have realized ρ as a half line spectral measure of this system also. Now we can use the transformations from Theorem 5.19 to pass to a modified version H_2 of H_1, which will have the same m function as our original canonical system H. So $H_2 = H$. But in fact we can say more. The transformations from Theorem 5.19 do not affect the de Branges space $B(E)$ because they either leave E unchanged (transformation of type (1)), or they change E by an $\mathrm{SL}(2, \mathbb{R})$ matrix as in Theorem 4.5 (transformation of type (2)). So $B(E) \equiv B(E_L)$ as de Branges spaces, not just after restricting (we already knew that both spaces equal $L^2(\mathbb{R}, \rho)$ then). □

As promised, we can now complete the proof of Theorem 5.11. The only case still left open arises when $B(E^{(2)}) \equiv L^2(\mathbb{R}, \rho)$ and $B(E^{(1)}) \subseteq L^2(\mathbb{R}, \rho)$ (superscripts will help avoid confusion with subscripts referring to regular points, to be introduced in a moment). We must show that then $B(E^{(1)}) \subseteq B(E^{(2)})$. To do this, we simply realize $B(E^{(2)}) = B(E_L)$ as the de Branges space of some canonical system, and then we repeat the arguments from above: by Theorem 4.10, $\rho = \rho_L^{(\beta)}$ is a spectral measure of this canonical system, and we can make it a half line spectral measure by adding a singular half line to our original system. Now Theorem 5.20 shows that $B(E^{(1)}) = B(E_x)$ for some regular $x > 0$, and the largest regular point is L, so $x \le L$, and thus $B(E_x) = B(E^{(1)})$ is isometrically contained in $B(E_L) = B(E^{(2)})$, as desired, by Theorem 4.8(a), (c).

Having worked with regular de Branges spaces so extensively, it feels appropriate now to also formulate a result that relates these to canonical systems, so let me do this. Everything is immediately clear from what we did already. I will map back and forth between canonical systems and de Branges *functions;* if you wanted to focus on the *spaces* instead, then there would be the additional option of conjugating H as in transformation (2) of Theorem 5.19 in part (b).

Theorem 5.21. (a) *Let E be a regular de Branges function with $E(0) = 1$. Then there are $L > 0$ and $H \in \mathcal{C}(L)$ such that $E(z) = E_L(z)$.*
(b) *Two canonical systems $H_j \in \mathcal{C}(L_j)$ with (say) $b := L_2 - L_1 \ge 0$ satisfy $E_{L_1}(z) = E_{L_2}(z)$ if and only if*

$$H_2(x) = \begin{cases} P_{e_2}, & 0 < x < b, \\ H_1(x-b), & x > b. \end{cases}$$

Proof. Part (a) follows from Theorem 5.2 since $(A, C)^t$ can be the first column of a $T \in TM$. Part (b) is immediate from Theorem 4.22(b). □

5.7 Notes

Section 5.1. Almost all results of this chapter are due to de Branges. The one-to-one correspondence between canonical systems and generalized Herglotz functions, however, is not stated explicitly in his work, and in fact m functions are never mentioned in [17]; they do not seem to be an object of interest from de Branges's point of view. I am fairly confident, though, that de Branges was for example well aware of what is Theorem 5.19 here, which says essentially the same thing as Theorem 5.1. In any event, the uniqueness of $H \in \mathcal{C}$ for a given m function was first made explicit in [67].

A completely different analysis of the existence part of Theorem 5.2 is given in [52].

Section 5.2. These metrics and their compactness properties inevitably play a large role in the proofs of the existence parts and are thus also used, at least implicitly, in de Branges's work; they are a rather standard tool, too. I give them a more detailed and explicit treatment here. They are also important in other investigations, and we will use them extensively in Chapter 7.

The reader interested in a direct proof of the compactness of $TM(L)$ could take a look at Lemma 5 of [14] and also Section 8.1 of [46], where the relevant tools are provided.

Section 5.4. The version of de Branges's ordering theorem given here is fully appropriate for our purposes, but it is not the most general. The statement also holds in certain situations for not necessarily regular spaces, and this has a similar proof, with only a moderate amount of additional complications. See [17, 59] for this. The proof I give here follows de Branges's own proof very closely, as do in fact all the published proofs, of which there are not that many. There is a small gap in de Branges's proof of his theorem in both [16] and [17]: when adapting Carleman's method, he ignores the possibility that we could have $L(t) = 2\pi$. This gap has been filled in [59]. Furthermore, most of the published proofs only establish what is our Proposition 5.13 and not Theorem 5.11 itself, and they do not always realize that there is a difference between the two. Some kind of argument is certainly needed, and it cannot be too soft as there are trivial examples of (non-regular) de Branges spaces that are isometrically contained in $L^2(\mathbb{R}, \rho)$ after restriction, but not in one another; one could make all spaces one-dimensional in such an example. The elegant and slick argument I present here that establishes Proposition 5.18 is due to de Branges.

In the main part of the argument, I have also gratefully incorporated some simplifying ideas from the presentation in [23], which become available in our situation, when the de Branges spaces are regular (though the proof in this reference has both gaps).

Carleman's method [11] (Lemmas 5.15 and 5.17 here) is a surprisingly powerful tool that has also worked its magic in other situations to produce sharp results, and some very accessible discussions of this may be found in [30, 63]. The original application was to what is now called the Denjoy–Carleman–Ahlfors theorem on asymptotic values of entire functions. Note also that in our situation, nothing less than the sharp constant in Wirtinger's inequality will do here; the slightest loss would immediately render the whole estimate of Q^2 useless.

Section 5.5. At first, it seems that it should be easy at this point to produce an $H \in \mathcal{C}(L)$ that realizes a given $T \in TM(L)$. After all, we can realize any Herglotz function, so why do we not take something like $m = -B/A$ and use Theorem 5.1 to find an H that has this as its m function; then do we not have every right to expect that this will have a transfer matrix with the correct first row somewhere? Once this is guaranteed, then we can remove any remaining discrepancy by hand, as in the actual proof.

One can make this work, and de Branges presents various arguments along these lines, but it is not easy at all because there is a rather obstinate technical problem that needs to be addressed: the H constructed above is really expected to end with a singular half line (if we had the theorem already, then it would be clear that $-B/A$ is the m function of a problem on a bounded interval), but this is not easy to show, and if we do not know it, then things simply do not get off the ground. In his treatments, de Branges uses the conjugate map $F \mapsto \widetilde{F}$ that I briefly mentioned in my notes to Chapter 4.

I use a new device here to address this issue, or rather bypass it completely: I instead realize the measure $d\rho(t) = dt/(\pi|E|^2)$, which is guaranteed to overshoot what we want since $L^2(\mathbb{R}, \rho)$ is certainly larger than $B(E)$, so that then $B(E)$ must sit at some finite L. This is technically very convenient and shortens the argument quite a bit, at the expense though of relying on the ordering theorem in this part also.

Section 5.6. The basic method employed here, to fit de Branges spaces into a chain associated with a canonical system by waiting for the moment when the inclusion behavior switches, is standard and in fact the only one that has ever been used for these general uniqueness proofs. Obviously, it depends crucially on de Branges's ordering theorem.

It might be interesting to investigate if the ordering theorem and thus the rather deep complex analysis behind it as well as the very notion itself of a de Branges space are really necessary to prove the uniqueness statements of Theorems 5.1, 5.2. It appears that these issues are also closely related to the invariant subspaces of certain Volterra operators on the spaces L^2_H, such as

$$(Vf)(x) = J \int_x^L H(t)f(t)\, dt,$$

and these questions could perhaps be analyzed directly. See also [27].

6 Some applications

6.1 Generalized and classical moment problems and the Nevanlinna parametrization

In this chapter, I will present various applications of the results of the previous chapter, in a rather unsystematic fashion. As our first topic, we return to the Weyl disks

$$\mathcal{D}(L;z) = \{T^{-1}(L;z)q : q \in \overline{\mathbb{C}^+}\}.$$

The one-to-one correspondence between canonical systems and Herglotz functions now gives us a crystal clear understanding of these: $\mathcal{D}(L;z)$ collects all those values that the m function can still take at z, given $H(x)$ on $(0,L)$. The following result is an expanded version of this now obvious remark.

Theorem 6.1. *Let $H_0 \in \mathcal{C}(L)$, denote its transfer matrix by $T(z) = T(L;z)$ and let $\mathcal{D}(L;z)$ be the corresponding Weyl disk. Then the following statements are equivalent for a canonical system $H \in \mathcal{C}$:*
(a) *$H(x) = H_0(x)$ on $0 < x < L$;*
(b) *$m(z;H) \in \mathcal{D}(L;z)$ for all $z \in \mathbb{C}^+$;*
(c) *$m(z;H) = T^{-1}(z)q(z)$ for some $q \in \mathcal{F}$.*

Proof. Recall that if m_L denotes the m function of the problem on (L,∞), then $m(z) = T^{-1}(L;z)m_L(z)$, as we see from the definition of $m(z) = f(0,z)$ as the value of the L^2_H solution at $x = 0$.

It is then obvious that (a) implies (b) and (c). Conversely, if (c) is assumed, then there is an $H_q \in \mathcal{C}$ with $q(z) = m(z;H_q)$, and thus

$$H_1(x) = \begin{cases} H_0(x), & 0 < x < L, \\ H_q(x-L), & x > L \end{cases}$$

also has $m(z;H)$ as its m function. The uniqueness of H now gives (a).

Finally, as we already observed earlier, in the context of Theorem 4.31, (b) and (c) are equivalent: if (b) is assumed, then we have the formula from part (c) for some function $q : \mathbb{C}^+ \to \overline{\mathbb{C}^+}$, and then it also follows that $q(z) = T(z)m(z;H)$ is holomorphic, so $q \in \mathcal{F}$. The converse is obvious. □

A slightly different viewpoint is even more illuminating. We now focus on the associated measures rather than the m functions themselves. It is useful to introduce some terminology, which will emphasize the analogy to the classical moment problem, that this, the problem of finding measures μ with prescribed *moments* $m_j = \int t^j\, d\mu(t)$.

https://doi.org/10.1515/9783110563238-006

Definition 6.1. Let $B(E)$ be a de Branges space, and let μ be a Borel measure on $\mathbb{R}$. We say that μ solves the *generalized moment problem (GMP)* $B(E)$ if $B(E)$ is isometrically contained in $L^2(\mathbb{R}, \mu)$.

This generalizes the classical moment problem: if $B(E)$ is a space of polynomials, then μ will solve the GMP $B(E)$ if and only if it has the correct moments. We will discuss this more explicitly below, after stating and proving the fundamental result on GMPs.

Condition (b) from Theorem 6.1, of lying in the corresponding Weyl disk for all $z \in \mathbb{C}^+$, essentially characterizes the solutions of a GMP $B(E)$. I said *essentially* because there is a small technical point involved here, and this is best taken care of by making the following definition.

Definition 6.2. Let $T \in TM$, $T \not\equiv 1$, and $q \in \mathcal{F}$. Denote the associated canonical systems by $H_T \in \mathcal{C}(L)$ and $H_q \in \mathcal{C}$, respectively. We call q *inadmissible* for T if $(0, L)$ ends with a singular interval of H_T and $(0, \infty)$ starts with a singular interval of H_q, and the types agree.

Obviously, this is a very special situation. For example, if H_T does not end with a singular interval, then all $q \in \mathcal{F}$ are admissible for T.

It was convenient to state the condition in terms of the canonical systems, but it is of course not necessary to construct these explicitly to decide whether or not a given $q \in \mathcal{F}$ is admissible for a $T \in TM$: by Theorem 4.34 and its proof, H_q starts with a singular interval if and only if $b(q) > 0$ or $q(iy) = a + O(1/y)$, and the type can also be identified from these data. Similarly, Theorem 4.36 lets us detect singular intervals of H_T at the right endpoint and their types by looking at T only, if we apply it to T and its transformed versions $R_\beta T R_{-\beta}$. Or we could use Theorem 4.13 to characterize the existence of a final singular interval in a different way.

Theorem 6.2. *Let $E = A - iC$ be a regular de Branges function with associated transfer matrix $T \in TM$ and let μ be a Borel measure on $\mathbb{R}$. Then μ solves the GMP $B(E)$ if and only if there is an admissible $q \in \mathcal{F}$ such that the Herglotz function $M(z) = T^{-1}(z)q(z)$ has μ as its measure.*

Recall that T is not uniquely determined by E; the theorem will work for any $T \in TM$ that has $(A, C)^t$ as its first column. In fact, that much is clear already because what is not determined about T is the length of a possible initial singular interval of type e_2 of the associated H_T, but this only affects $b(M)$ and not its measure.

Proof. If μ is such that $B(E) \subseteq L^2(\mathbb{R}, \mu)$ isometrically, then $\int \frac{d\mu(t)}{1+t^2} < \infty$, by Lemma 5.12. Hence there are $H \in \mathcal{C}$ that have μ as their spectral measure, and for any such H, it then follows from Theorem 5.20 that $B(E) \equiv B(E_L)$ for some regular $L > 0$. We have some freedom in the choice of H here, as spelled out in Theorem 5.19, and we can use the transformation of type (2) to make $E(z) = E_L(z)$. This gives $T(z)$ and $T(L; z; H)$ the same first column, and now changing the length of a possible initial singular interval of type e_2 (corresponding to the transformation of type (1) from Theorem 5.19) will make $T(z) = T(L;z;H)$ (the value of L might actually be changed by this, but I will not reflect this in the notation).

As in the previous proof, we now have $m(z;H) = T^{-1}(z)q(z)$, with q being the m function of the problem on (L,∞). By construction, $m(z;H)$ has measure μ, and q is admissible for T because L was a regular point, and the transformations we applied to H do not change this.

Conversely, assume now that μ is the measure of $M = T^{-1}q$, for some admissible $q \in \mathcal{F}$. Consider the canonical systems $H_T \in \mathcal{C}(L)$, $H_q \in \mathcal{C}$ corresponding to T and q, respectively, and then let

$$H(x) = \begin{cases} H_T(x), & 0 < x < L, \\ H_q(x-L), & x > L. \end{cases}$$

Then $M(z) = m(z;H)$ is the m function of this problem, and $E(z) = E_L(z;H)$. Since q was admissible for T, we know that L is a regular point of H, and thus $B(E_L) \subseteq L^2(\mathbb{R},\mu)$ isometrically, by Theorem 4.8(b), (c). This conclusion is valid even if $(0,L)$ is a singular interval because we can still be sure then that it will not be of type e_2. □

This seemed rather straightforward, but of course the proof was short and easy only because we now have extremely powerful tools at our disposal. Theorem 6.2 can be viewed as a vast generalization of the Nevanlinna parametrization of the solutions of (classical) indeterminate moment problems, and this latter result is usually considered a rather substantial achievement in its own right.

Let us now discuss how this classical result becomes a special case of Theorem 6.2. This, too, is quite straightforward, but, unsurprisingly, it will depend on some basic facts about the classical moment problem. I do not want to discuss this material at great length here, so will assume some familiarity with it during this brief digression. There are many good introductions; for our needs here, [64] is particularly useful.

Let a set of (potential) moments $m_j \in \mathbb{R}$, $j \geq 0$, be given. We impose the well-known necessary and sufficient condition for these numbers to actually be the moments of some (necessarily finite) measure μ that is not supported by a finite set, namely, we assume that

$$\sum_{j,k\geq 0} m_{j+k}\overline{p_j}p_k > 0 \tag{6.1}$$

for all finitely supported sequences (p_j), not identically equal to zero. It is of course obvious that this condition is necessary for the existence of μ because the sum computes $\int |p(t)|^2\,d\mu(t)$, $p(t) = \sum p_j t^j$, if we have a measure μ with the required moments. The converse is a classical result in the theory of moments.

Now given such a sequence m_j, the classical moment problem

$$\int_{-\infty}^{\infty} t^j\,d\mu(t) = m_j, \quad j \geq 0, \tag{6.2}$$

is called *determinate* if there is exactly one solution μ, and it is called *indeterminate* otherwise. A solution of (6.2) is defined in the obvious way as a measure μ with $\int(1+t^{2n})\,d\mu(t) < \infty$ for all $n \geq 1$ that satisfies (6.2).

The Nevanlinna parametrization describes the set of solutions to an indeterminate moment problem, so let us now assume that we are given such an indeterminate moment sequence m_j. Since we can compute integrals of polynomials with respect to μ using the moments only, we can apply the Gram–Schmidt procedure to the monomials $1, t, t^2, \ldots$ to form orthogonal polynomials p_j, and these will satisfy a Jacobi difference equation (though not with bounded coefficients this time). This Jacobi operator will be in the limit circle case at infinity; this condition is in fact equivalent to the moment problem being indeterminate. The spectral measures coming from the m function of the Jacobi operator (with arbitrary boundary condition at infinity) solve the moment problem.

Now we rewrite this Jacobi difference equation as a canonical system, in exactly the same way as discussed in Section 5.3. Recall that this canonical system will consist of a succession of singular intervals, of lengths (using the notation from that section) $L_n = p_n^2 + q_n^2$. Recall further that p_n, q_n were solutions to the Jacobi difference equation at $z = 0$. Thus, by the limit circle property, $L := \sum L_n < \infty$. This was to be expected, since a limit circle Jacobi matrix presumably cannot very well correspond to a limit point canonical system.

So we have now rewritten the Jacobi operator as a canonical system $H \in \mathcal{C}(L)$. The interval $(0, L)$ consists of a succession of singular intervals, and there are infinitely many of these, and they accumulate at L. The first singular interval is not of type e_2 (it is of type e_1, as we saw in Section 5.3). Since the set R of regular points accumulates at L, we have

$$B(E_L) = \overline{\bigcup_{x \in R, x < L} B(E_x)}. \tag{6.3}$$

Moreover, for each $x \in R$, $x < L$, the corresponding space $B(E_x)$ is finite-dimensional and contains exactly the polynomials of at most a certain degree. This we see as follows: $(0, x)$ consists of finitely many singular intervals, and let us say there are n of them. Since the initial singular interval is not of type e_2, we then know that the space $\overline{D(\mathcal{S})}$ from Theorem 3.19 and thus also $L^2(\mathbb{R}, \rho)$ is n-dimensional, if we avoid the exceptional boundary condition at x. Here, $\mathcal{S}$ denotes a self-adjoint realization of the canonical system on $(0, x)$, and ρ is the corresponding spectral measure. So the spectrum $\sigma(\mathcal{S})$ has exactly n points in it, and it is given by the solutions t of $e_\beta^* J u(x, t) = 0$. This is true for all boundary conditions β, with one exception. The transfer matrix $T(x; z)$ is a product of n matrices of the type $1 + zL_j JP_{\alpha_j}$, so is a polynomial of degree at most n, and what we just observed about the spectrum now implies that its first column $u(x, z)$ must be of degree exactly n. It follows that $E_x(z) = u_1(x, z) - iu_2(x, z)$ also is a polynomial of degree n. All zeros of E_x are in $\mathbb{C}^-$, and therefore $p \in B(E_x)$ for all polynomials p with $\deg p \leq n - 1$. This is already an n-dimensional space, and

$\dim B(E_x) = n$ since $B(E_x) \equiv L^2(\mathbb{R}, \rho)$, so $B(E_x) = \mathcal{P}_{n-1}$, the space of polynomials of degree at most $n-1$.

Now the spectral measures ρ of the problems on $(0, x)$ that were discussed above are exactly the spectral measures of the corresponding truncated (finite) Jacobi matrices. This follows as in Theorem 5.10 from the fact that the m functions (of the Jacobi matrix and the canonical system) agree. As I mentioned already, the spectral measures of the (truncated) Jacobi matrices solve the (truncated) moment problem. So if ρ is the spectral measure of a canonical system on $(0, x)$, with non-exceptional boundary condition at x, as above, then

$$\int_{-\infty}^{\infty} t^j \, d\rho(t) = m_j, \quad j = 0, 1, \ldots, 2n-2. \tag{6.4}$$

Since $B(E_x) \equiv L^2(\mathbb{R}, \rho)$, this says that solving the GMP $B(E_x)$ is the same thing as solving the truncated classical moment problem (6.4).

Theorem 6.3. *Consider the canonical system $H \in \mathcal{C}(L)$ that was constructed above. Then the measure μ solves the GMP $B(E_L)$ if and only if it solves the classical moment problem* (6.2).

Proof. This summarizes what we just did. We also use the fact that solving the GMP $B(E_L)$ is the same as solving the GMPs $B(E_x)$ for all $x \in R$, $x < L$, and this follows from (6.3). □

We have set the stage, and the classical Nevanlinna parametrization of the solutions of (6.2) can now be obtained effortlessly. We denote the transfer matrix of the canonical system from Theorem 6.3 by $T(z) = T(L; z)$.

Corollary 6.4 (Nevanlinna). *Let μ be a Borel measure on $\mathbb{R}$. Then μ solves the classical moment problem* (6.2) *if and only if μ is the measure of the Herglotz function $M(z) = T^{-1}(z)q(z)$, for some $q \in \mathcal{F}$.*

Proof. This is an immediate consequence of Theorem 6.2. It only remains to note that since $(0, L)$ does not end with a singular interval, all $q \in \mathcal{F}$ are admissible for T. □

One could even say that this way of viewing things is slightly superior to the classical version of the result in that we have $T(z)$ available immediately and do not need to construct it via a limiting process. We also obtain additional insight into the nature of the solution set because we have a second, equivalent description: the measures μ that solve the moment problem are exactly the half line spectral measures of coefficient functions $H \in \mathcal{C}$ that agree with the H constructed above on $(0, L)$.

As a by-product of this whole analysis, we also obtain another criterion for the determinacy of a moment problem. Suppose that moments satisfying (6.1) are given. This lets us define a scalar product $[\cdot, \cdot]$ on the space $\mathcal{P}$ of polynomials by this formula.

Theorem 6.5. *The moment problem is indeterminate if and only if the completion of* $(\mathcal{P}, [\cdot,\cdot])$ *is a regular de Branges space.*

Or, to state this condition very explicitly and carefully, what we are asking for is the existence of a regular de Branges space $B(E) \supseteq \mathcal{P}$ such that the identity map $\mathcal{P} \to B(E)$, $p \mapsto p$, has an extension to a unitary map from the completion onto $B(E)$. This is of course the same as asking that $\mathcal{P}$ is isometrically contained in $B(E)$ and dense in this space.

Proof. We just discussed at some length that this completion equals $B(E_L)$ if the moment problem is indeterminate. Conversely, if the space equals some regular de Branges space $B(E)$, then solving the original classical moment problem (6.2) is the same as solving the GMP $B(E)$, and this GMP has many solutions by Theorem 6.2. □

Let us now return to GMPs. We can add more detail to the picture painted by Theorem 6.2. The correspondence between solutions μ of the GMP and admissible generalized Herglotz functions q will turn out to be a bijection. Moreover, in much the same way as discussed in Section 5.2, this map is continuous in both directions if certain natural metrics are used. We do not necessarily have compact spaces here, though, due to the possible presence of inadmissible generalized Herglotz functions.

For $q \in \mathcal{F}$, we of course use the metric on that space that we already have, and as for the measures μ, we recall that convergence of Herglotz functions implies the weak-$*$ convergence of the measures $d\nu(t) = \frac{d\rho(t)}{1+t^2} + b\delta_\infty$ of the modified Herglotz representation; see formula (3.9) of Chapter 3. This suggests to use a metric d with the property that $d(\mu_n, \mu) \to 0$ if and only if

$$\int_{-\infty}^{\infty} f(t)\frac{d\mu_n(t)}{1+t^2} \to \int_{-\infty}^{\infty} f(t)\frac{d\mu(t)}{1+t^2} \tag{6.5}$$

for all $f \in C(\mathbb{R}_\infty)$. We can construct such a d in the same way as discussed in Section 5.2. With these preparations in place, everything now works as hoped for.

Theorem 6.6. *In the situation of Theorem 6.2, the map* $q \mapsto \mu$ *from admissible generalized Herglotz functions to solutions of the GMP is a bijection. Both the map and its inverse are continuous with respect to the metrics just discussed.*

Proof. That the map is surjective onto the solutions of the GMP is part of what Theorem 6.2 says. We must still check that it is injective as well. If we have admissible $q_1, q_2 \in \mathcal{F}$ that lead to the same μ, then the Herglotz functions $M_j = T^{-1}q_j$ satisfy $M_1 = M_2 + a + bz$, for some $a, b \in \mathbb{R}$. This means that M_1, M_2 are the m functions of two canonical systems that are related to one another by the transformations of Theorem 5.19, corresponding to these two parameters a, b, but of course both coefficient functions start with the same $H = H_T$ on an initial interval, so $b = 0$ and then $a = 0$ as well (or we could apply Theorem 5.21, or use the argument based on the Weyl disks $\mathcal{D}(iy)$). So $q_1 = q_2$.

The continuity of the direct map $q \mapsto \mu$ is obvious from what we just discussed: if $q_n \to q$, then also $M_n \to M$ locally uniformly, so

$$b(M_n)\delta_\infty + \frac{d\mu_n(t)}{1+t^2} \to b(M)\delta_\infty + \frac{d\mu(t)}{1+t^2}$$

in weak-$*$ sense as measures on $\mathbb{R}_\infty$. Here, $b(M)$ of any $M = T^{-1}q$ is already determined by T and does not depend on q, so $b(M_n) = b(M)$, and thus it also follows that $d(\mu_n, \mu) \to 0$.

Conversely, suppose now that $d(\mu_n, \mu) \to 0$, and μ_n, μ all solve the GMP $B(E)$. Denote the associated admissible generalized Herglotz functions by q_n, q. By the compactness of $\mathcal{F}$, we can pass to a subsequence (but written as if it were the original sequence, as usual) such that $q_n \to p$ for some $p \in \mathcal{F}$, possibly inadmissible. Then also $T^{-1}q_n \to T^{-1}p$, but from our assumption, we also know that these functions can only converge to a Herglotz function F with $b(F) = b(T^{-1}q_n)$ (recall that this is independent of n) and with measure μ. In other words, $T^{-1}p = a + T^{-1}q$. However, this now implies $p = q$ as in the first part of the proof, so q is the only possible limit point of the sequence q_n and thus $q_n \to q$, as claimed. □

This proof, in its last part, has actually shown slightly more than stated, and while this is a technical point, it might be of some interest also: suppose now that, instead of (6.5), we only have this convergence with an extra term $+bf(\infty)$ on the right-hand side. In other words, some or all of the mass of $d\mu_n/(1+t^2)$ is allowed to leak out to infinity, and it is then not contained in $d\mu/(1+t^2)$. Then it still follows that $q_n \to q$, by the same proof, except that now, we can only infer that $T^{-1}p = a + bz + T^{-1}q$, but this still gives the desired conclusion $p = q$.

So both versions of (6.5), the original one and the modification by $bf(\infty)$ just described, are equivalent for solutions of the GMP, and mass can actually not leak out to infinity.

Corollary 6.7. *The space of solutions to the GMP $B(E)$, with the metric described above, is compact if and only if all $q \in \mathcal{F}$ are admissible.*

Proof. By Theorem 6.6, the space is homeomorphic to the space of admissible $q \in \mathcal{F}$. Since an inadmissible q can be always approximated by admissible functions, it is clear that this latter space is compact if and only if there are no inadmissible $q \in \mathcal{F}$. □

6.2 Existence of moments

Having discussed moments so extensively in the previous section, it is now natural to wonder what the condition that $\int(1+t^{2n})\,d\rho(t) < \infty$ on a spectral measure means for the corresponding canonical systems. We will obtain a more elegant formulation of these results if we first get the following easy special case out of the way.

Theorem 6.8. *The spectral measure ρ is finitely supported if and only if the set R of regular points is finite.*

Proof. The spectral measure ρ is finitely supported if and only if $L^2(\mathbb{R},\rho)$ is finite-dimensional, and this happens if and only if $\overline{D(S)}$ is finite-dimensional, and now the description of this space from Theorem 3.19 makes it clear that this is equivalent to R being finite. □

To avoid having to distinguish cases, as in Theorem 4.34 (which is the special case $n = 0$ of the result I am about to formulate), it is convenient to introduce one more piece of terminology: for a canonical system $H \in \mathcal{C}$, not a single singular half line, define its *reduced* version H_r by removing a possible initial singular interval of type e_2. More explicitly, we set $H_r(x) = H(x+c)$, with $c = c(H) \geq 0$, and $c(H)$ is of course defined as in Section 4.7 as the length of the initial singular interval of type e_2 if there is one, and $c(H) = 0$ otherwise.

Theorem 6.9. *Suppose that the spectral measure ρ of $H \in \mathcal{C}$ is not supported by a finite set. Then*

$$\int_{-\infty}^{\infty} (1 + t^{2n})\, d\rho(t) < \infty \tag{6.6}$$

if and only if H_r starts with $n + 1$ singular intervals.

Proof. We already proved one direction, in the previous section in the discussion leading up to Theorem 6.3: if H_r starts with $n + 1$ singular intervals on $(0, L)$, then $B(E_L^{(r)})$ contains the polynomials of degree at most n. Since $E_L^{(r)} = E_{L+c}$, the same is true for a de Branges space of the non-reduced system. Since $B(E_{L+c})$ is isometrically contained in $L^2(\mathbb{R},\rho)$, it follows that $1, t^n \in L^2(\mathbb{R},\rho)$ also, as claimed.

So now assume that (6.6) holds. We will use the argument from the proof of Theorem 4.34, in a more elaborate version. Notice, first of all, that the property of H_r of starting with $n + 1$ singular intervals is invariant under the transformations from Theorem 5.19, and of course (6.6) is not affected either since the spectral measure does not change, so we may apply these transformations freely. At the level of the m functions, they add or subtract terms of the form $a + bz$. So we may now assume that the m function of the reduced system is of the form

$$m(z) = \int_{-\infty}^{\infty} \frac{d\rho(t)}{t - z}.$$

By expanding

$$\frac{1}{t - z} = -\sum_{j=0}^{2n-1} \frac{t^j}{z^{j+1}} - \frac{t^{2n}}{z^{2n+1}} \frac{1}{1 - t/z},$$

we see that

$$m(iy) = -\sum_{j=0}^{2n-1} c_j(iy)^{-j-1} + O(y^{-2n-1}), \quad y \to \infty, \tag{6.7}$$

with $c_j = \int t^j \, d\rho(t)$.

In particular, $m(iy) = -c_0/(iy) + o(1/y)$, so

$$\frac{-1}{m(iy)} = \frac{iy}{c_0} + o(y),$$

and now we see from Theorem 4.34 and its proof that the modified H_r starts with a singular interval of type e_1 and of length $L_1 = 1/c_0$.

Now we look at the m function m_1 of this problem on (L_1, ∞). We have

$$m_1(z) = (1 + zL_1JP_{e_1})m(z) = \begin{pmatrix} 1 & 0 \\ L_1z & 1 \end{pmatrix} m(z) = \frac{m(z)}{zL_1m(z) + 1},$$

or, equivalently,

$$-\frac{1}{m_1(z)} = -\frac{1}{m(z)} - L_1z.$$

From the expansion (6.7) of m, we also obtain the corresponding expansion

$$-\frac{1}{m(iy)} = \frac{iy}{c_0} + d_1 + d_2(iy)^{-1} + \cdots + d_{2n-1}(iy)^{-2n+2} + O(y^{-2n+1}),$$

with coefficients $d_j \in \mathbb{R}$ that could in principle be found from the moments c_j, but we do not need to know them here. Right now, the important thing is that then also

$$-\frac{1}{m_1(iy)} = d_1 + d_2(iy)^{-1} + \cdots + d_{2n-2}(iy)^{-2n+2} + O(y^{-2n+1}). \tag{6.8}$$

This function $-1/m_1 = Jm_1$ is the m function of $-JH_1(x)J$, if we write H_1 for the coefficient function that corresponds to m_1 itself. Note that this transformation, too, from H_1 to $-JH_1J$, preserves all singular intervals and only changes their types. Now we can use a second transformation of the canonical system on (L_1, ∞), of the type discussed above, to make $d_1 = 0$ in (6.8), and then we are in exactly the same situation that we were in before, with m instead of $-1/m_1$; see (6.7) again. It follows that (L_1, ∞) also starts with a singular interval.

We need to address one issue here: if $d_2 = 0$ in (6.8), then the argument will not work as intended (for example, the formula for the length of the new singular interval is $L_2 = 1/d_2$). However, in that case, the generalized Herglotz function $F = -1/m_1 - d_1$ would satisfy $F(iy) = o(1/y)$, and this is impossible for a (genuine) Herglotz function. Thus $F \equiv 0$, but that means that (L_1, ∞) is a singular half line, which is also impossible since the set R of regular points is infinite by assumption.

We can now repeat this step from m to m_1 with $-1/m_1 - d_1$ as the new starting point, and the method will deliver singular intervals as long as there are still terms left from the expansion, for a total of $n+1$ singular intervals, as required; the production of a new singular interval costs us two terms from the asymptotic expansion.

One more technical point needs clarification here: in the final step, we will have an m function with asymptotics $-1/m_n(iy) = O(1/y)$, rather than the more precise information that $-y/m_n(iy)$ approaches a limit, as before. It is therefore important that this is good enough to conclude that there is an initial singular interval of type e_1, but this works: $yF(iy) = O(1)$ for a Herglotz function F will imply that $\lim_{y\to\infty} yF(iy)$ exists, and this follows because $G = -1/F$ also is a Herglotz function, and for any Herglotz function, $\lim_{y\to\infty} G(iy)/y$ exists (and computes $ib(G)$). □

Corollary 6.10. *Suppose that the spectral measure ρ is compactly supported. Then $R \cap (0, L)$ is finite for all $L > 0$.*

Proof. Such a measure has moments of all orders, so Theorem 6.9 gives us an infinite sequence of consecutive singular intervals $(0, L_1)$, (L_1, L_2), etc. (in the non-trivial case when ρ is not finitely supported). We must show that $L_n \to \infty$ here. If this is false, so $L_n \to L < \infty$, then, as discussed in the previous section, the GMP $B(E_L)$ is a classical indeterminate moment problem, and it has ρ as a solution. It is well known that indeterminate moment problems cannot have compactly supported solutions. For example, this follows from Carleman's criterion; see again [64] for more on this. □

The converse of the corollary is false: there are limit point Jacobi matrices with unbounded coefficients, and such a matrix, being an unbounded operator, cannot have a spectral measure with compact support. Thus such a Jacobi difference equation provides a counterexample when rewritten as a canonical system.

6.3 Diagonal canonical systems

In this quick and easy section, we characterize the titular diagonal canonical systems and a related subclass of canonical systems in terms of their spectral data.

If $T(x; z)$, $0 \le x \le L$, is the transfer matrix of an $H \in \mathcal{C}(L)$, then the transformed matrix

$$T_1(x; z) = IT(x; -z)I, \quad I = \begin{pmatrix} 1 & 0 \\ 0 & -1 \end{pmatrix}$$

solves $JT_1' = -zH_1T_1$, $T_1(0) = 1$, with $H_1(x) = IH(x)I$, as we see from a calculation. If we write $H = \left(\begin{smallmatrix} \alpha & \beta \\ \beta & \gamma \end{smallmatrix}\right)$, then

$$H_1(x) = \begin{pmatrix} \alpha(x) & -\beta(x) \\ -\beta(x) & \gamma(x) \end{pmatrix},$$

so this has the following consequences.

Theorem 6.11. *Let $H \in \mathcal{C}(L)$, and denote the entries of $T = T(L;z)$ by A, B, C, D, as usual, and $E = E_L = A - iC$. The following statements are equivalent:*
(a) *$H(x)$ is a diagonal matrix for $0 < x < L$;*
(b) *$A(z), D(z)$ are even functions and $B(z), C(z)$ are odd;*
(c) *$E^{\#}(z) = E(-z)$.*

Proof. The matrix $H(x)$ is diagonal if and only if $H_1(x) = H(x)$, so the one-to-one correspondence between canonical systems and transfer matrices shows that (a) is equivalent to $T_1 = T$, and if this is written out, then it becomes condition (b). The implication from (b) to (c) is trivial.

Regarding the de Branges functions $E = E_L$, the calculation we did shows that (for general $H \in \mathcal{C}(L)$) the de Branges function E_1 of H_1 is related to the one of H by $E_1(z) = E^{\#}(-z)$. Hence, if (c) is assumed, it follows that H_1 and H have the same de Branges function. Obviously, if one of these coefficient functions starts with a singular interval of type e_2, then so does the other, and the lengths agree. Thus Theorem 5.21(b) now shows that $H_1 = H$, that is, (a) holds. □

To derive analogous results on half line problems, we first look at what the transformation from H to H_1 does to the m function. The above calculation also clarifies this at once.

Theorem 6.12. *Let $H \in \mathcal{C}$. Then $m_1(z) = -m^{\#}(-z)$.*

Notice that in this formula, m is evaluated at $-\overline{z}$, which will lie in $\mathbb{C}^+$ if z was taken from $\mathbb{C}^+$, so everything is well defined here.

Proof. Let $z \in \mathbb{C}^+$ and write $w = -\overline{z} \in \mathbb{C}^+$. Since $T = T^{\#}$, we then have

$$T_1(x;z)\begin{pmatrix} -m^{\#}(-z) \\ 1 \end{pmatrix} = -IT(x;-z)\begin{pmatrix} m^{\#}(-z) \\ 1 \end{pmatrix}$$
$$= -I\overline{T(x;w)\begin{pmatrix} m(w) \\ 1 \end{pmatrix}}.$$

This lies in $L^2_{H_1}(0,\infty)$, and this property identifies $-m^{\#}(-z)$ as $m_1(z)$. □

Theorem 6.13. *Let $H \in \mathcal{C}$, denote its m function by $m(z)$, and let ρ be the associated measure. The following statements are equivalent:*
(a) *$H(x)$ is diagonal for $x > 0$;*
(b) *$m(z) = -m^{\#}(-z)$;*
(c) *$\operatorname{Re} m(iy) = 0$ for all $y > 0$;*
(d) *$\operatorname{Re} m(i) = 0$, and ρ is even in the sense that $\rho(A) = \rho(-A)$ for every Borel set $A \subseteq \mathbb{R}$.*

Proof. As in the proof of Theorem 6.11, (a) is equivalent to $H_1 = H$, which is equivalent to (b) by Theorem 6.12. The equivalence of (b) and (c) is trivial; to deduce (b) from (c), look at the holomorphic function $m(z) + m^{\#}(-z)$. Finally, condition (d) is also easily related to (b), (c). If (b) is assumed, then certainly Re $m(i) = 0$, and if $f \in C(\mathbb{R})$ is compactly supported, then

$$\int_{-\infty}^{\infty} f(t)\, d\rho(t) = \frac{1}{\pi} \lim_{y\to 0+} \int_{-\infty}^{\infty} f(t) \operatorname{Im} m(t+iy)\, dt = \int_{-\infty}^{\infty} f(-t)\, d\rho(t),$$

as required; the second equality follows by noting that $\operatorname{Im} m(t+iy)$ is an even function of t.

Conversely, if (d) is assumed, then use the Herglotz representation in the modified version

$$m(iy) = biy + \int_{-\infty}^{\infty} \frac{1-tiy}{t-iy} \frac{d\rho(t)}{1+t^2};$$

in general, there would be an extra constant a on the right-hand side, but we see from the formula itself that this equals $a = \operatorname{Re} m(i)$, so is absent here. Now take real parts and observe that the resulting integrand is an odd function of t, so (c) follows. □

An analogous set of results can be obtained if we perform an additional transformation and define $T_2(x;z) = -JT_1(x;z)J$, with the T_1 from above. This also lies in TM, and the corresponding coefficient function is given by $H_2(x) = -JH_1(x)J$, or, if we write it out,

$$H_2(x) = \begin{pmatrix} \gamma(x) & \beta(x) \\ \beta(x) & \alpha(x) \end{pmatrix}.$$

This time, we have $H_2(x) = H(x)$ if and only if $\alpha(x) = \gamma(x) = 1/2$. Thus repeating the above arguments now produces the following results (notations are also as above).

Theorem 6.14. *Let $H \in \mathcal{C}(L)$. Then $\alpha(x) = \gamma(x) = 1/2$ on $0 < x < L$ if and only if $A(z) = D(-z)$, $B(z) = C(-z)$.*

In the analog of Theorem 6.13, we will make use of what is usually called the *Krein function* of a Herglotz function F. To define it for a given F, we take the holomorphic logarithm $\log F(z)$, $z \in \mathbb{C}^+$, with $0 < \operatorname{Im} \log F < \pi$, and then let

$$\xi(t) = \frac{1}{\pi} \lim_{y\to 0+} \operatorname{Im} \log F(t+iy).$$

The limit will exist for almost every $t \in \mathbb{R}$, and $0 \le \xi(t) \le 1$. The measure associated with the Herglotz function $\log F$ is $\xi\, dt$, and $b(\log F) = 0$.

Theorem 6.15. *Let $H \in \mathcal{C}$. The following statements are equivalent:*
(a) *$\alpha(x) = \gamma(x) = 1/2$ for $x > 0$;*
(b) *$m(z) = 1/m^{\#}(-z)$;*
(c) *$|m(iy)| = 1$ for all $y > 0$;*
(d) *$|m(i)| = 1$, and the Krein function $\xi(t)$ of m is even.*

Proof. Since this is both very straightforward and completely analogous to what we did above, I will only discuss the one new aspect here, the use of the Krein function. So I will only show that (c) is equivalent to (d).

This, too, is obvious because (c) of the current theorem is the same statement as (c) of Theorem 6.13 for $F = \log m$, and, similarly, (d) of Theorem 6.15 is the same as (d) of Theorem 6.13 for the corresponding data of F, which are $\log |m(i)|$ and $\xi \, dt$. □

It is also clear from these last observations (if it was not before) that if m is as in Theorem 6.15, then $F = \log m$ will be the m function of a diagonal canonical system. Of course, we are not obtaining all diagonal canonical systems in this way; we obtain precisely those whose spectral measure is purely absolutely continuous with density ≤ 1.

6.4 Dirac systems

We continue to investigate diagonal canonical systems, and we make the additional assumptions that $H \in AC$ and $\det H(x) \neq 0$ for all $x > 0$. It will then actually be convenient to normalize not the trace but the determinant of $H(x)$ and choose the variable x such that $\det H(x) = 1$. This has the advantage that x has immediate spectral theoretic meaning as the type of the transfer matrix, by Theorem 4.26. So, to summarize, we are now interested in canonical systems with coefficient functions of the form

$$H(x) = \begin{pmatrix} a(x) & 0 \\ 0 & a^{-1}(x) \end{pmatrix},$$

and here $a > 0$ is assumed to be an absolutely continuous function of $x \geq 0$.

These canonical systems can be rewritten as Dirac equations. To do this, consider $y(x) = H(x)^{1/2} u(x)$. Then a quick calculation shows that u solves $Ju' = -zHu$ if and only if y solves

$$Jy'(x) = W(x) \begin{pmatrix} 0 & 1 \\ 1 & 0 \end{pmatrix} y(x) - zy(x), \quad W(x) = \frac{a'(x)}{2a(x)}. \tag{6.9}$$

Our assumptions on a make W a locally integrable function. Conversely, if a Dirac system (6.9) with $W \in L^1_{\text{loc}}$ is given, then taking these steps in reverse produces a diagonal canonical system whose coefficient function, which we may now take as $a(x) = \exp(2 \int_0^x W(t)\, dt)$, satisfies the same assumptions as above.

The Dirac system formulation (6.9) has one big advantage over the canonical system, and its exploitation will be the main theme of this section: a rather precise asymptotic analysis is possible. This is already quite plausible on an intuitive level since for large z, the first term on the right-hand side looks unimportant compared to the contribution zy; in fact, most of the classical equations have specific asymptotics built into them in this way, whereas for a general canonical system, there is no obvious approximate simplification that takes place at large z, and indeed what could be said in this generality required abstract complex analytic tools, as we saw in Chapter 4.

Our analysis will focus on the de Branges function

$$u_1(L,z) - iu_2(L,z) = a^{-1/2}(L)y_1(L,z) - ia^{1/2}(L)y_2(L,z).$$

The factors $a^{\pm 1/2}(L)$ are not relevant: we can use Theorem 4.5 to transform them away without changing the de Branges space, and indeed a more convenient choice of de Branges function is

$$E_L(z) = y_1(L,z) - iy_2(L,z), \quad y(0,z) = a(0)^{1/2}\begin{pmatrix} 1 \\ 0 \end{pmatrix}. \tag{6.10}$$

Here, the factor $a(0)^{1/2}$ has the trivial effect of rescaling the norm on $B(E_L)$, without changing this space otherwise, so we will disregard it and simply assume in the sequel that $a(0) = 1$.

We want to analyze (6.9) by treating W as a perturbation, and to do this, we introduce

$$Q(x,z) = \begin{pmatrix} e^{izx} & e^{-izx} \\ -ie^{izx} & ie^{-izx} \end{pmatrix}.$$

Then Q solves the equation with $W = 0$, that is, $JQ' = -zQ$. We now define $v = v(x,z)$ by writing $y = Qv$. Another quick calculation then shows that v solves

$$v' = W(x)\begin{pmatrix} 0 & e^{-2izx} \\ e^{2izx} & 0 \end{pmatrix} v, \quad v(0,z) = \frac{1}{2}\begin{pmatrix} 1 \\ 1 \end{pmatrix},$$

or, if we write this in integrated form,

$$v(x,z) = \frac{1}{2}\begin{pmatrix} 1 \\ 1 \end{pmatrix} + \int_0^x W(t)\begin{pmatrix} 0 & e^{-2izt} \\ e^{2izt} & 0 \end{pmatrix} v(t,z)\, dt. \tag{6.11}$$

This equation may now be solved by iteration, by putting

$$v_0 = \frac{1}{2}\begin{pmatrix} 1 \\ 1 \end{pmatrix}, \quad v_{n+1} = \frac{1}{2}\begin{pmatrix} 1 \\ 1 \end{pmatrix} + \int_0^x W(t)\begin{pmatrix} 0 & e^{-2izt} \\ e^{2izt} & 0 \end{pmatrix} v_n(t)\, dt.$$

It will be convenient to write the (formal, at this point) solution as a series

$$v(x,z) = \frac{1}{2}\begin{pmatrix} 1 \\ 1 \end{pmatrix} + \sum_{n\ge 1} f_n(x,z), \tag{6.12}$$

with

$$f_n(x,z) = \frac{1}{2}\int_0^x dt_1\, W(t_1)\int_0^{t_1} dt_2\, W(t_2)\ldots\int_0^{t_{n-1}} dt_n\, W(t_n)\begin{pmatrix} e^{-2iz(t_1-t_2+t_3-\cdots\pm t_n)} \\ e^{2iz(t_1-t_2+t_3-\cdots\pm t_n)} \end{pmatrix}.$$

It is now obvious from this that (6.12) does converge to a solution, uniformly on $0 \le x \le L$ and on $z \in K$ for any compact subset $K \subseteq \mathbb{C}$. Indeed, we then have the estimates

$$\|f_n(x,z)\| \le \frac{\|W\|_{L^1(0,L)}^n}{\sqrt{2}n!} e^{2|z|L},$$

which are immediate from (6.12) once we observe that in the region of integration the argument

$$s = t_1 - t_2 + t_3 - \cdots \pm t_n \tag{6.13}$$

of the exponential function satisfies $0 \le s \le L$.

We can say much more about the structure of v. Since both v and the terms f_n from its series representation have the property that $v_2(x,z) = v_1(x,-z)$, we can focus on one of its components, and I will take the second one, v_2. Of course, we then also only need the second components of the f_n, which I will denote by $g_n(x,z)$. Recall also that our goal was to analyze the de Branges function E_L from (6.10), which in terms of v can be written as

$$E_L(z) = 2v_2(L,z)e^{-iLz}. \tag{6.14}$$

By making $2s$ from (6.13) one of the integration variables, we see that each g_n is the Fourier transform

$$g_n(L,z) = \widehat{h_n}(z) = \int h_n(x)e^{izx}\, dx$$

of some function $h_n \in L^1(0,2L)$. Moreover, the series $\sum h_n$ converges in $L^1(0,2L)$, by the same estimates as above, and thus also $2v_2 - 1 = \widehat{h}$, for some $h \in L^1(0,2L)$. If we now use this in (6.14), then we see that we have proved the following.

Theorem 6.16. *Let $W \in L^1(0,L)$. Then the de Branges function from* (6.10) *of the corresponding Dirac system satisfies*

$$E_L(z) = e^{-iLz} + \widehat{f}(z)$$

for some real valued $f \in L^1(-L,L)$.

This is already quite interesting, but a variant of this statement will be even more useful, to clarify the nature of the de Branges space $B(E_L)$.

Theorem 6.17. *In the situation of Theorem* 6.16, *we also have*

$$\frac{1}{|E_L(t)|^2} - 1 = \widehat{\phi}(t) \quad (t \in \mathbb{R}),$$

for some even, real valued function $\phi \in L^1(\mathbb{R})$.

To derive this from Theorem 6.16, we will need Wiener's lemma, in the following version.

Lemma 6.18 (Wiener). *Suppose that $f \in L^1(\mathbb{R})$, $\widehat{f}(t) \neq -1$ for all $t \in \mathbb{R}$. Then*

$$\frac{1}{1+\widehat{f}} = 1 + \widehat{g}$$

for some $g \in L^1(\mathbb{R})$.

Proof. We will adapt the well-known Banach algebra proof of the original version of Wiener's lemma to our setting. The required tools are discussed in more detail in, for example, [61, Chapter 11].

Consider the Banach algebra

$$\mathcal{A} = \{c + \widehat{h}(t) : c \in \mathbb{C}, h \in L^1(\mathbb{R})\},$$

with the algebraic operations defined pointwise and $\|c + \widehat{f}\| = |c| + \|f\|_{L^1}$. Or, put differently, we are dealing with L^1 with its convolution product, and we have adjoined a unit ("δ") to this algebra. The functions from $\mathcal{A}$ are continuous on $\mathbb{R}_\infty$, and the complex homomorphisms of $\mathcal{A}$ are precisely given by the point evaluations

$$\phi_t(c + \widehat{h}) = c + \widehat{h}(t), \quad t \in \mathbb{R}_\infty.$$

So if f is as in the lemma, then $\phi_t(1 + \widehat{f}) \neq 0$ for all complex homomorphisms; for $t = \infty$, we see this from the Riemann–Lebesgue lemma. It follows that $1+\widehat{f}$ is invertible in $\mathcal{A}$, and the inverse must be of the form $1 + \widehat{h}$, for the same reason, because $(1 + \widehat{f})(\infty) = 1$. □

Proof of Theorem 6.17. We return to the formula we originally obtained for E_L; we have

$$E_L(t) = e^{-iLt}(1 + \widehat{g}(t))$$

for some $g \in L^1(0, 2L)$. So

$$\frac{1}{|E_L(t)|^2} - 1 = \frac{1}{|1 + \widehat{g}(t)|^2} - 1 = -\frac{\widehat{h}(t)}{1 + \widehat{h}(t)},$$

with

$$\widehat{h} = \widehat{g} + \widehat{g_r} + \widehat{g}\widehat{g_r}, \quad g_r(x) := g(-x).$$

The reflected function g_r makes an appearance here because $\overline{\widehat{g}(t)} = \widehat{g_r}(t)$ for real valued g. Since $L^1(\mathbb{R})$ is closed under the convolution product, which becomes pointwise multiplication after taking the Fourier transforms, we have $h \in L^1(\mathbb{R})$, and thus Wiener's lemma applies. Note that we do have $\widehat{h} \neq -1$ since $E_L(t) \neq 0$ for $t \in \mathbb{R}$.

Thus $1/|E|^2 - 1 = -\widehat{h}(1 + \widehat{k})$ for some $k \in L^1(\mathbb{R})$, and this is the Fourier transform of some $L^1(\mathbb{R})$ function ϕ, as claimed.

The additional statements about ϕ are obvious because $1/|E|^2 - 1$ is real valued and even, by Theorem 6.11(b), and thus so will be its Fourier transform ϕ. □

If $W = 0$, then $\phi = 0$ also here and thus $E_L(z) = E_0(z) = e^{-iLz}$. We already discussed this de Branges function in Section 4.2. The corresponding de Branges space is the Paley–Wiener space

$$PW_L = \{F(z) = \widehat{f}(z) : f \in L^2(-L, L)\}$$

and $\|F\|^2_{B(E_0)} = 2\|f\|_2^2$ by Plancherel's identity.

Theorem 6.19. *For any $W \in L^1(0, L)$, we have $B(E_L) = PW_L$ as sets, and*

$$\|F\|^2_{B(E_L)} = 2\|f\|^2 + 2\langle f, \phi * f\rangle, \tag{6.15}$$

for $F = \widehat{f} \in PW_L$, with the function $\phi \in L^1(\mathbb{R})$ from Theorem 6.17.

The convolution $\phi * f$, with $\phi \in L^1(\mathbb{R})$, $f \in L^2(-L, L)$, defines a function from $L^2(\mathbb{R})$, by Young's inequality (or, simpler perhaps, by recalling that $\widehat{\phi}$ is bounded). Of course, when we evaluate the scalar product with $f \in L^2(-L, L)$ in (6.15), only its values on $(-L, L)$ matter. As a consequence, $\|F\|_{B(E_L)}$ is determined by (and determines) the restriction of ϕ to $(-2L, 2L)$. Theorem 6.20 below will state the same fact in a different way.

If we denote this convolution operator by $C_\phi f = \phi * f$, then we can write (6.15) as

$$\|F\|^2 = 2\langle f, (1 + C_\phi)f\rangle. \tag{6.16}$$

The operator C_ϕ is self-adjoint and compact, and $1 + C_\phi$ must be strictly positive, or (6.16) would not define a norm. So the eigenvalues are bounded away from zero, and we now see that $B(E_L)$ and $B(E_0) = PW_L$ are not only equal as sets, but these spaces also have equivalent norms. The effect of the potential $W \in L^1(0, L)$ is limited to a relatively small distortion of the norm.

Proof. By Theorem 6.16, in the version

$$E_L(z) = e^{-iLz}(1 + \widehat{g}(z)) = E_0(z)(1 + \widehat{g}(z)),$$

with $g \in L^1(0, 2L)$, we have $C_1|E_0(z)| \le |E_L(z)| \le C_2|E_0(z)|$ on the closed upper half plane $\operatorname{Im} z \ge 0$, for some constants $C_1, C_2 > 0$. Here we again use $\widehat{g}(x + iy) \to 0$ as $|x| \to \infty$, by the Riemann–Lebesgue lemma, and of course the continuous function $\widehat{g}$ cannot take the value -1, so $|1 + \widehat{g}| \ge \delta > 0$ on $\mathbb{C}^+$, as required.

It is now obvious from this estimate that $B(E_L) = B(E_0) = PW_L$ as sets. The formula for the norm then follows from Theorem 6.17, by writing $1/|E_L|^2 = 1 + \widehat{\phi}$ and taking (inverse) Fourier transforms. □

Theorem 6.20. *Let μ be a positive Borel measure on $\mathbb{R}$ that satisfies $\int_{-\infty}^{\infty} \frac{d\mu(t)}{1+t^2} < \infty$. Define the signed measure $d\sigma(t) = d\mu(t) - dt/\pi$. Then μ solves the GMP $B(E_L)$ if and only $\widehat{\sigma} = 2\phi$ on $(-2L, 2L)$.*

In this version, the statement is succinct and also immediately plausible from Theorem 6.19, but in fact we must be careful here to ensure that everything makes rigorous sense since the Fourier transform of a measure is not guaranteed to be a function and thus must not, in general, be evaluated pointwise.

To address these technical issues, we observe, first of all, that our assumption on μ does make this measure a tempered distribution, so $\widehat{\mu}$ and $\widehat{\sigma}$ are well defined as distributions. These can be restricted to an open set, simply by only applying them to test functions supported by this set, and the statement of the theorem is to be taken in this sense. Of course, once we have done all this, it then turns out that $\widehat{\sigma}$ is an integrable function on $(-2L, 2L)$; $\widehat{\mu}$ has the extra contribution 2δ.

Proof. Suppose first that μ solves the GMP $B(E_L)$. Theorem 6.19 then shows that

$$2\langle f, \phi * f\rangle = \int_{-\infty}^{\infty} |\widehat{f}(t)|^2 \, d\sigma(t) \tag{6.17}$$

for all $f \in C_0^{\infty}(-L, L)$.

The Fourier transform of the tempered distribution σ is defined by the condition that $(\widehat{\sigma}, g) = (\sigma, \widehat{g})$ for all test functions g. If we take specifically a g of the form $g = f * f_r$, with $f \in C_0^{\infty}(-L, L)$ and $f_r(x) = \overline{f(-x)}$, then $\widehat{g} = |\widehat{f}|^2$. Moreover, we can rewrite the left-hand side of (6.17) as

$$\langle f, \phi * f\rangle = \int_{-L}^{L} dx\, \overline{f(x)} \int_{-2L}^{2L} dt\, \phi(t) f(x-t) = \int_{-2L}^{2L} \phi(t)(f * f_r)(-t)\, dt,$$

and since ϕ is even, this equals $\int \phi(t) g(t)\, dt$, and thus (6.17) becomes

$$2(\phi, g) = (\widehat{\sigma}, g); \tag{6.18}$$

this holds for all $g = f * f_r, f \in C_0^{\infty}(-L, L)$.

We want to conclude that $\widehat{\sigma} = 2\phi$ from this, so we must make sure that this set of functions $g = f * f_r$, with $f \in C_0^{\infty}(-L, L)$, is sufficiently rich to determine a distribution on $(-2L, 2L)$. This follows because if we have (6.18) for these g, then we also obtain it for $g = f * h_r$, with $f, h \in C_0^{\infty}(-L, L)$, by polarization, and then also for linear combinations of such functions. It is now easy to see that this set of functions is dense in the appropriate test function space: to approximate a general $k \in C_0^{\infty}(-2L, 2L)$, we can break it into pieces $k = k_1 + k_2 + k_3$ whose supports are of diameter $< 2L$. For such a function k_j, fix an $a \in (-L, L)$ such that $k_j(t-a)$ is supported by $(-L, L)$, and then k_j can be approximated in the required fashion by taking convolutions with an approximation to the identity localized near a.

If, conversely, $\widehat{\sigma} = 2\phi$ on $(-2L, 2L)$, then we certainly have (6.18) and thus also (6.17), for $f \in C_0^\infty(-L, L)$, by taking the above steps in reverse order. We are not quite done yet because we need this statement or its equivalent version $\|F\|_{B(E_L)} = \|F\|_{L^2(\mu)}$ for arbitrary $f \in L^2(-L, L)$. To obtain this, approximate such an f by $f_n \in C_0^\infty(-L, L)$ in $L^2(-L, L)$. As I pointed out above, following the statement of Theorem 6.19, this norm is comparable with the one on $B(E_L)$, so $\|F_n - F\|_{B(E_L)} \to 0$ as well. The existence of reproducing kernels (or a simple direct argument) now gives that $F_n(t) \to F(t)$ pointwise, and thus Fatou's lemma shows that

$$\begin{aligned}\|F\|_{L^2(\mu)} &\le \lim_{n\to\infty}\left(\int_{-\infty}^{\infty} |F_n(t)|^2\, d\mu(t)\right)^{1/2}\\ &= \lim_{n\to\infty} \|F_n\|_{B(E_L)} = \|F\|_{B(E_L)}.\end{aligned}$$

This inequality holds for all $F \in B(E_L)$, and if we apply it to $F_n - F$, then it shows that $F_n \to F$ also in $L^2(\mu)$, and this finally gives $\|F\|_{L^2(\mu)} = \|F\|_{B(E_L)}$, as required. □

6.5 Notes

Section 6.1. The major result of this section, Theorem 6.2, is due to de Branges, in various versions. He gives it a slightly different formulation; my notion of inadmissible $q \in \mathcal{F}$ seems to be new (though quite obvious of course). Instead, de Branges discusses explicitly how the measures obtained from inadmissible qs fail to integrate correctly on a one-dimensional subspace of $B(E)$. This information could also be extracted quite easily from Theorem 6.2 as stated and our previous results, but I have not made it explicit here.

Nevanlinna's original result, Corollary 6.4, is from [49]. Quite a few additional things could and have been said about it and its ramifications, but I again refer to the literature on the classical moment problem for this.

That the classical moment problem can be embedded into the spectral theory of canonical systems is well known, and two relevant references are [32, 58]. In fact, much of de Branges's work was originally motivated (as far as one can tell) by this connection. I have not seen the determinacy criterion Theorem 6.5 stated anywhere, but of course it is an immediate consequence of the material developed here.

Some authors define spectral measures ρ of canonical systems $H \in \mathcal{C}$ by just asking that U maps $\overline{D(\mathcal{S})}$ isometrically into $L^2(\mathbb{R}, \rho)$, without ever mentioning the multiplication operator M_t in $L^2(\mathbb{R}, \rho)$. This works, as is shown by the following result (which follows at once from the material of this section).

Theorem 6.21. *Let $H \in \mathcal{C}$ and suppose that* $(0, \infty)$ *does not end with a singular half line. Then the spectral measure that is obtained from the m function is the unique measure that solves all GMPs $B(E_L)$, $L \in \mathbb{R}$.*

That said, this alternative definition has little to recommend it. As their name suggests, spectral measures are supposed to be useful in the analysis of the spectral properties of $\mathcal{S}$, and they are, but the reason for this is that they realize the operator as a multiplication operator, so dropping this property from the definition is a bit like defining a car as something that can be stored in a garage.

Section 6.2. Both the question answered by Theorem 6.9 and its proof are so natural that the result can hardly be new, but I have not found it in the literature.

Section 6.3. Diagonal canonical systems are equivalent to Krein strings. For more on this connection, see [23, 34, 59, 71].

Section 6.4. This section contains some previously unpublished material. However, all the methods and ideas used here are very well known. The function ϕ is a key object in inverse spectral theory done in Gelfand–Levitan style. The whole analysis could be carried much further, and one could set up a one-to-one correspondence between potentials $W \in L^1(0, L)$ and $\phi \in L^1(-2L, 2L)$, where these latter functions must of course satisfy the condition that was discussed after the statement of Theorem 6.19: $1 + C_\phi$ must be a positive operator. A lengthy discussion of the inverse spectral theory of Schrödinger operators from this point of view is given in [53].

7 The absolutely continuous spectrum

7.1 Twisted shifts and Toda maps

In this chapter, we will study the absolutely continuous spectrum of a canonical system on $\mathbb{R}$ (mainly) or on $(0,\infty)$. The recurring theme will be the idea to analyze this canonical system H not in isolation, but by looking at limit points $K = \lim F_n(H)$ that are obtained by applying certain maps to the original system and then taking limits with respect to the metrics that we already introduced in Section 5.2.

This first section will provide a general framework for these investigations. We will work mainly with trace normed whole line problems, and we allow arbitrary coefficient functions H of this type. So $H(x) = P_\alpha$, corresponding to a single singular interval equal to all of $\mathbb{R}$, is now possible. We denote the collection of these coefficient functions by $\mathcal{C}(\mathbb{R})$, and we put a metric d on this space of the type discussed in Section 5.2. We can either build d from scratch, in the same way as before, or, more conveniently perhaps, we can simply set

$$d(H,K) = d_+(H_+,K_+) + d_-(H_-,K_-),$$

with $H_\pm, K_\pm$ denoting the restrictions to the half lines $(-\infty,0)$, $(0,\infty)$, and then $d_\pm$ can be literally the same metrics as the ones constructed in Section 5.2.

Then $(\mathcal{C}(\mathbb{R}),d)$ becomes a compact metric space. Of course, this would not have worked if we had not included the coefficient functions $H \equiv P_\alpha$, so this is really more or less forced on us and not an optional feature. The map $\mathcal{C}(\mathbb{R}) \to \mathcal{F}^2$, $H \mapsto (m_-,m_+)$ is a homeomorphism onto the space of pairs of generalized Herglotz functions.

Now consider the shift map

$$(t \cdot H)(x) = H(x+t)$$

on $\mathcal{C}(\mathbb{R})$. The notation I chose here emphasizes the fact that this defines an action of the group $(\mathbb{R},+)$ on $\mathcal{C}(\mathbb{R})$. We can now ask ourselves what the corresponding map on $\mathcal{F}^2$ is that is induced by this, and the answer is already clear: m functions are updated by letting the transfer matrix act on them, and this was immediate from the fact that $m_\pm(z) = \pm f_\pm(0,z)$, since T updates $f_\pm$. More explicitly, we have

$$m_\pm(z;t\cdot H) = \pm T(t;z;H)\big(\pm m_\pm(z;H)\big).$$

Limit points of shifts of H will play a very important role in the study of the absolutely continuous spectrum, but in fact shifts are not a sufficiently general class of maps for canonical systems. To motivate this remark, consider a (whole line) Schrödinger equation

$$-y''(x) + V(x)y(x) = zy(x), \tag{7.1}$$

https://doi.org/10.1515/9783110563238-007

with potential $V \in L^1_{\text{loc}}(\mathbb{R})$, and rewrite it as a canonical system, with coefficient function

$$H = \begin{pmatrix} p^2 & pq \\ pq & q^2 \end{pmatrix}, \tag{7.2}$$

as discussed in more detail in Section 1.3. (This H is not trace normed, but this will not affect the point I am trying to make right now.) Recall that p, q were the solutions of (7.1) with $z = 0$ and with the initial values $p'(0) = q(0) = 1$, $p(0) = q'(0) = 0$.

Now replace $V(x)$ with its shifted version $(t \cdot V)(x) = V(x + t)$, and also write this Schrödinger equation as a canonical system, in the same way. The corresponding coefficient function will usually not be $H(x + t)$ because $p(x + t)$, $q(x + t)$ no longer have the required initial values at $x = 0$. The correct way to transform H is as follows: the new solutions p_t, q_t that we need can be obtained as

$$\begin{pmatrix} p_t'(x) & q_t'(x) \\ p_t(x) & q_t(x) \end{pmatrix} = T_0(x + t)T_0^{-1}(t),$$

and here (as in Section 1.3)

$$T_0(x) = \begin{pmatrix} p'(x) & q'(x) \\ p(x) & q(x) \end{pmatrix}$$

denotes the transfer matrix of the Schrödinger equation, written in matrix form, at $z = 0$. If we now use p_t, q_t as the correct substitutes for p, q in (7.2), then we obtain the transformed coefficient function $t \cdot H$, corresponding to the shifted potential $t \cdot V$, as

$$(t \cdot H)(x) = T_0^{-1t}(t)H(x + t)T_0^{-1}(t).$$

This motivates the following definition.

Definition 7.1. Let $H \in \mathcal{C}(\mathbb{R})$. A *twisted shift* of H is a new coefficient function $H_2 \in \mathcal{C}(\mathbb{R})$ which is obtained as the trace normed version of

$$H_1(x) = A^{-1t}H(x + t)A^{-1},$$

for some $t \in \mathbb{R}$, $A \in SL(2, \mathbb{R})$. We also refer to $t \in \mathbb{R}$ as the *length* of the twisted shift.

By what we observed above about plain shifts and Theorem 3.20 (and its analog for m_-), if $K \in \mathcal{C}(\mathbb{R})$ is a twisted shift of $H \in \mathcal{C}(\mathbb{R})$, then

$$m_\pm(z; K) = \pm AT(t; z; H)(\pm m_\pm(z; H)).$$

Maps of this type preserve the spectral properties. This is of course trivial for shifts, and it is also completely unsurprising for twisted shifts, but the statement also holds for a much larger class of maps, which we now introduce.

Definition 7.2. A *Toda map* is a transformation $H_1 \mapsto H_2$ between canonical systems $H_j \in \mathcal{C}(\mathbb{R})$ of the type

$$m_\pm(z; H_2) = \pm S(z)(\pm m_\pm(z; H_1)),$$

for some entire matrix function $S : \mathbb{C} \to \mathrm{SL}(2, \mathbb{C})$ with $S = S^\#$.

So twisted shifts are special Toda maps, but this latter notion is much wider: we have considerably relaxed our requirements on $S(z)$. The terminology alludes to the fact that if a Jacobi matrix is evolved by a Toda flow, then the corresponding maps induced by the flow will be of this type. This also holds for other evolution equations from the classical integrable hierarchies such as KdV type flows on Schrödinger operators. Conversely, these evolutions are the main source of examples for Toda maps that are not just twisted shifts.

It is almost slightly misleading to call this a map since given an $S(z)$ as above and a canonical system $H \in \mathcal{C}(\mathbb{R})$, there is of course no guarantee that the two new functions $\pm S(z)(\pm m_\pm(z; H))$ will be generalized Herglotz functions. For example, in the extreme case when $H \equiv P_\alpha$ is a singular line, then $-m_- = m_+ = a \in \mathbb{R}_\infty$, and the transformation defines a new canonical system only if both $S(z)a$ and $-S(z)a$ are generalized Herglotz functions, and this happens if and only if $S(z)a = b \in \mathbb{R}_\infty$. Thus such an H can be mapped exactly to the canonical systems $H \equiv P_\beta$ of the same type by a Toda map.

The only Toda maps that can be applied to any $H \in \mathcal{C}(\mathbb{R})$ are the ones corresponding to a constant $S(z) = S \in \mathrm{SL}(2, \mathbb{R})$. Now S just acts by a fixed automorphism of $\mathbb{C}^\pm$ on $\pm m_\pm(z)$; we have used these transformations quite frequently already.

Toda maps also have an easy description when viewed as transformations of the matrix valued Herglotz functions

$$M(z) = \frac{-1}{2(m_+ + m_-)} \begin{pmatrix} -2m_+m_- & m_+ - m_- \\ m_+ - m_- & 2 \end{pmatrix}$$

that we use for the spectral representation of the whole line problem, as discussed in Section 3.7. Since these are undefined if $H \equiv P_\alpha$, we must exclude these canonical systems here.

Theorem 7.1. *Suppose that $H_j \in \mathcal{C}(\mathbb{R})$, $H_j \not\equiv P_\alpha$, are related by a Toda map $H_1 \mapsto H_2$, induced by $S(z)$. Then*

$$M_2(z) = S(z) M_1(z) S^t(z).$$

Proof. A brute force calculation would of course (eventually) establish this, but a slightly less computational proof is also possible. Notice that MJ has the half line m functions as eigenvectors. More precisely, we have

$$2M(z)J \begin{pmatrix} \pm m_\pm(z) \\ 1 \end{pmatrix} = \mp \begin{pmatrix} \pm m_\pm(z) \\ 1 \end{pmatrix}, \tag{7.3}$$

at least if we assume that $m_\pm \not\equiv \infty$. Moreover, since $H \not\equiv P_\alpha$, the two vectors $(\pm m_\pm(z), 1)^t$ are linearly independent for any $z \in \mathbb{C}^+$, so these evaluations characterize the matrix $M(z)$. Now recall the identity $JS = S^{-1t}J$, which is valid for any $S \in SL(2, \mathbb{C})$, fix $z \in \mathbb{C}^+$, and write

$$c_\pm(z)S(z)\begin{pmatrix} \pm m_\pm^{(1)}(z) \\ 1 \end{pmatrix} = \begin{pmatrix} \pm m_\pm^{(2)}(z) \\ 1 \end{pmatrix},$$

with $c_\pm \in \mathbb{C}$, $c_\pm \neq 0$; this again requires $m_\pm^{(j)}$ to be distinct from infinity. Under these assumptions, we compute

$$\begin{aligned} 2SM_1S^tJ\begin{pmatrix} \pm m_\pm^{(2)} \\ 1 \end{pmatrix} &= 2c_\pm SM_1S^tJS\begin{pmatrix} \pm m_\pm^{(1)} \\ 1 \end{pmatrix} \\ &= 2c_\pm SM_1J\begin{pmatrix} \pm m_\pm^{(1)} \\ 1 \end{pmatrix} = \mp c_\pm S\begin{pmatrix} \pm m_\pm^{(1)} \\ 1 \end{pmatrix} \\ &= \mp\begin{pmatrix} \pm m_\pm^{(2)} \\ 1 \end{pmatrix}. \end{aligned}$$

As explained, this implies the asserted formula for $M_2(z)$.

Note that this argument was purely algebraic, and it did not make use of the Herglotz property of our functions. We showed that if two numbers $\pm m_\pm \in \mathbb{C}$, $m_+ \neq -m_-$, are transformed by an $S \in SL(2, \mathbb{C})$, then the corresponding matrices M transform as asserted in the theorem. Thus we now also obtain the originally excluded cases, when one of the functions is identically equal to infinity, by a limiting argument. (Of course, it is also possible to adapt the calculation itself to these cases; for example, if $m_- = \infty$, then the corresponding part of (7.3) becomes $2MJe_2 = -e_2$.) □

Theorem 7.2. *Suppose that $H_1, H_2 \in \mathcal{C}(\mathbb{R})$ are related by a Toda map. Then the self-adjoint operators S_1, S_2 generated by these canonical systems are unitarily equivalent.*

Proof. Everything is already clear in the trivial case when H_1 or H_2 is a single singular line: then the other canonical system must be of this type also, as we saw, and both relations are purely multi-valued, and $\overline{D(\mathcal{S}_j)} = 0$.

In the other case, we have the matrix valued Herglotz functions M_j and their associated measures ρ_j available, and we must now show that the two multiplication operators M_t in $L^2(\mathbb{R}, \rho_j)$ are unitarily equivalent. This is the same as showing that ρ_1, ρ_2 are equivalent measures, including multiplicity. More specifically, we must show that both measures have the same null sets and the ranks of the densities of the absolutely continuous parts agree; recall that the singular parts can never have multiplicity greater than one, by Theorem 3.22.

All this is a quick consequence of Theorem 7.1. Let $F : \mathbb{R} \to \mathbb{C}^2$ be a compactly supported continuous function. We have

$$\int_{-\infty}^{\infty} F^*(t)\, d\rho_j(t)F(t) = \frac{1}{\pi}\lim_{y\to 0+}\int_{-\infty}^{\infty} F^*(t)\mathrm{Im}\, M_j(t+iy)F(t)\, dt$$

and, by Theorem 7.1,

$$\operatorname{Im} M_2(t+iy) = S(t)\operatorname{Im} M_1(t+iy)S^t(t) + O(y\|M_1(t+iy)\|),$$

since $S(z)$ is entire and real on the real line; the constant implicit in the error term can be taken to be independent of $0 < y \le 1$ and t from the support of F. For any Herglotz function P and any $R > 0$, we have

$$\lim_{y\to 0+} \int_{-R}^{R} y|P(t+iy)|\,dt = 0; \tag{7.4}$$

this follows comfortably from the Herglotz representation and does not require delicate estimates. If we work with the modified version with a finite measure ν on $\mathbb{R}_\infty$, then what we need to look at here is the double integral

$$y\int_{-R}^{R} dx \int_{\mathbb{R}_\infty} d\nu(t) \left|\frac{1+t(x+iy)}{t-x-iy}\right|; \tag{7.5}$$

in our intended application, we have a matrix valued Herglotz function, but of course we can control this by estimating the diagonal entries, so it suffices to discuss scalar Herglotz functions in this step. We now carry out the x integration first, so write (7.5) as

$$\int_{\mathbb{R}_\infty} f(t;y)\,d\nu(t), \quad f(t;y) = y\int_{-R}^{R} \left|\frac{1+t(x+iy)}{t-x-iy}\right| dx.$$

It is then easy to show that $f(t;y)$ stays bounded on $t \in \mathbb{R}_\infty$, $0 < y \le 1$, and $f(t;y) \to 0$ as $y \to 0+$, by considering separately the cases $|t| \le 2R$ (say) and $|t| > 2R$. Dominated convergence now gives (7.4).

This whole argument has shown that

$$d\rho_2(t) = S(t)\,d\rho_1(t)S^t(t),$$

and what we wanted to establish about these measures is now obvious from this identity. □

7.2 Convergence of Herglotz functions

We will soon need various equivalent descriptions of the convergence of elements of $(\mathcal{F}, d)$; these will refer to other quantities that we associate with Herglotz functions. The most obvious such assignment would be to relate the convergence of Herglotz functions to that of the measures. This is a standard result that is easy to prove, and we have been using it quite regularly already, in one form or another. Let us start out by giving it a precise formulation.

It will be convenient here to use the modified version of the Herglotz representation, so write

$$F(z) = a + \int_{\mathbb{R}_\infty} \frac{1+tz}{t-z}\, d\nu(t).$$

This formula also works for generalized Herglotz functions F as long as $F \not\equiv \infty$; in this case, $\nu = 0$, and a is the constant value of F.

Theorem 7.3. *Let $F_n, F \in \mathcal{F} \setminus \{\infty\}$. Then*
(a) *$d(F_n, F) \to 0$ if and only if $a_n \to a$ and $\nu_n \to \nu$, in weak-$*$ sense;*
(b) *$d(F_n, \infty) \to 0$ if and only if $|a_n| + \nu_n(\mathbb{R}_\infty) \to \infty$.*

Proof. It is a typical feature of these proofs (which we will encounter again several times below) that it is enough to establish one implication; the converse will then follow automatically from this, when combined with the compactness of $\mathcal{F}$.

Observe that $F(i) = a + i\nu(\mathbb{R}_\infty)$. So if $F_n \to F$, $F \not\equiv \infty$, then certainly $a_n \to a$. Moreover, ν_n is a bounded sequence of measures, so will converge in weak-$*$ sense on a subsequence to some (finite) Borel measure μ on $\mathbb{R}_\infty$. Then, by passing to the limit in the Herglotz representation formula, we see that (a, μ) work as the data for F, so by the uniqueness of these, it then follows that $\mu = \nu$ and the passage to a subsequence was unnecessary.

We have shown that if $F_n \to F \in \mathcal{F}$ (and $F \equiv \infty$ is allowed now), then we have the asserted convergence of (a_n, ν_n) as in part (a) or (b). If, conversely, this is assumed, then $F_n \to G \in \mathcal{F}$ on a subsequence by the compactness of $\mathcal{F}$, but then it follows, from what we already have, that (a_n, ν_n) must converge to the data of G, so $G = F$. □

The key object of the next result will be $\log F(t)$, the logarithm of the boundary values $F(t) \equiv \lim_{y\to 0+} F(t+iy)$ of the Herglotz function F. Recall that these exist almost everywhere with respect to Lebesgue measure. Moreover, if $F_1(t) = F_2(t)$ on a set $t \in A \subseteq \mathbb{R}$ of positive measure, then $F_1 \equiv F_2$. This holds more generally for functions from N, and in this setting, it is an immediate consequence of the fact that $\int |\log^- F(t)| \frac{dt}{1+t^2} < \infty$ if $F \not\equiv 0$. So in particular, we cannot have $F(t) = 0$ on a set of positive measure unless $F \equiv 0$, and thus $\log F(t)$ may indeed be defined at almost all $t \in \mathbb{R}$.

We have $F(t) \in \overline{\mathbb{C}^+}$, and we take the logarithm with $0 \le \operatorname{Im} \log F \le \pi$. In fact, as we already briefly discussed earlier, in Section 6.3, we can take a holomorphic logarithm of $F(z)$ itself, and $\log F(z)$ will then also be a Herglotz function. We can then view $\log F(t)$ also as the boundary value of this function; in other words, it does not matter in which order the two operations *take the logarithm* and *take the boundary values* are carried out.

Recall that we also introduced the *Krein function*

$$\xi(t) = \frac{1}{\pi} \operatorname{Im} \log F(t);$$

this satisfies $0 \le \xi \le 1$, and $\xi(t)\,dt$ is the measure of the Herglotz function $\log F$, and since clearly $b(\log F) = 0$, the two versions of its Herglotz representation are

$$\begin{aligned} \log F(z) &= C + \int_{-\infty}^{\infty} \left(\frac{1}{t-z} - \frac{t}{t^2+1} \right) \xi(t)\,dt \\ &= C + \int_{-\infty}^{\infty} \frac{1+tz}{t-z} \zeta(t)\,dt, \end{aligned} \tag{7.6}$$

with $\xi(t) = (1+t^2)\zeta(t)$.

Theorem 7.4. (a) *If $F \in \mathcal{F} \setminus \{0, \infty\}$, then $\log F(t) \in L^2(-R, R)$ for all $R > 0$.*

(b) *Let $F_n, F \in \mathcal{F} \setminus \{0, \infty\}$, and let $A \subseteq \mathbb{R}$ be a bounded Borel set of positive Lebesgue measure. Then $F_n \to F$ if and only if $\log F_n \to \log F$ weakly in $L^2(A)$.*

(c) *Let $F_n \in \mathcal{F} \setminus \{0, \infty\}$ and let $A \subseteq \mathbb{R}$ be a bounded Borel set of positive Lebesgue measure. Then:*

(i) *$F_n \to 0$ if and only if*

$$\int_A \log |F_n(t)|\,dt \to -\infty;$$

(ii) *$F_n \to \infty$ if and only if*

$$\int_A \log |F_n(t)|\,dt \to \infty.$$

Proof. We follow the same overall pattern as in the previous proof: once we have the direct implications of (b), (c), then the converses will just fall into place. We focus on the situation from part (b) for now, and assume that $F_n \to F$. Then also $\log F_n \to \log F$ locally uniformly, so, by Theorem 7.3, the data from (7.6) satisfy $C_n \to C$ and $\zeta_n\,dt \to \zeta\,dt$, and this latter convergence takes place in weak-$*$ sense, with $\zeta_n\,dt$, $\zeta\,dt$ viewed as measures on $\mathbb{R}_\infty$.

For this direct implication, it suffices to discuss the case when $A = (-R, R)$, for some $R > 0$. We then split the integral from (7.6) into two parts, corresponding to $|t| \le R+1$ and $|t| > R+1$. In this second part, the limit $z = x + iy \to x$ that must be taken to compute the boundary value $\log F(x)$ presents no problems: we can simply take the corresponding limit in the integrand. Furthermore, we will then have

$$\int_{|t|>R+1} \frac{1+tx}{t-x} \zeta_n(t)\,dt \to \int_{|t|>R+1} \frac{1+tx}{t-x} \zeta(t)\,dt, \tag{7.7}$$

weakly in $L^2(-R,R)$ as functions of x (and of course these functions, being bounded, do lie in $L^2(-R,R)$). This follows because $\zeta_n\,dt \to \zeta\,dt$ in weak-$*$ sense; in fact, it would not be hard to establish the much stronger statement (which we do not need here) that the convergence in (7.7) holds in pointwise sense and is uniform in $|x| \le R$.

To investigate the contributions coming from $|t| \le R+1$, we return to the alternative formula from (7.6) (the first line). Both integrals converge separately when restricted to $|t| \le R+1$, and clearly

$$\int_{-R-1}^{R+1} \frac{t}{t^2+1}\xi_n(t)\,dt \to \int_{-R-1}^{R+1} \frac{t}{t^2+1}\xi(t)\,dt,$$

by the weak-$*$ convergence $\xi_n\,dt \to \xi\,dt$; to address the small technical issue that the integrand $\chi_{(-R-1,R+1)}\cdots$ is not continuous, we use the uniform bound $\xi_n \le 1$.

It remains to analyze the functions

$$L_n(x) \equiv \lim_{y\to 0+} \int_{-R-1}^{R+1} \frac{\xi_n(t)\,dt}{t-x-iy}.$$

We must show that $L_n, L \in L^2(-R,R)$, and $L_n \to L$ weakly in this space. Now $\operatorname{Im} L_n = \pi\xi_n$ almost everywhere by the general fact that the imaginary part of the boundary value of a Herglotz function is (π times) the density of the absolutely continuous part of its measure. Then, since $0 \le \xi_n \le 1$, the weak-$*$ convergence also gives us the weak convergence of this sequence in $L^2(-R,R)$, by approximating a general test function $g \in L^2(-R,R)$ by continuous functions.

The real part of L_n can alternatively be computed as a *Hilbert transform* of $\chi_B(t)\xi_n(t)$, $B \equiv (-R-1, R+1)$, that is,

$$\operatorname{Re} L_n(x) = (H(\chi_B\xi_n))(x) \equiv \lim_{y\to 0+} \int_{|t-x|>y} \frac{\chi_B(t)\xi_n(t)}{t-x}\,dt.$$

The limit defining the principal value integral will exist for almost all x. These statements are not particularly easy to establish, but this is a standard piece of harmonic analysis and as such discussed in many books; see, for example, [35, Chapter VI.B].

The Hilbert transform H defines a bounded operator on $L^2(\mathbb{R})$, and $H^* = -H$. So, in particular, $\operatorname{Re} L_n \in L^2(-R,R)$ also, and we have now established part (a). Moreover, if $g \in L^2(A)$ is arbitrary, with $A = (-R,R)$, then

$$\begin{aligned}\langle g, \operatorname{Re} L_n\rangle_{L^2(A)} &= \langle \chi_A g, H(\chi_B\xi_n)\rangle_{L^2(\mathbb{R})} = -\langle H(\chi_A g), \xi_n\rangle_{L^2(B)} \\ &\to -\langle H(\chi_A g), \xi\rangle_{L^2(B)} = \langle g, \operatorname{Re} L\rangle_{L^2(A)}.\end{aligned}$$

Here, the convergence follows from the weak convergence $\xi_n \to \xi$ in $L^2(-R-1, R+1)$ that was observed earlier, and then the last equality is established by taking the first two steps in reverse.

We have now proved the direct implication of part (b), and then this direction of (c)(i) follows in the same way. The contribution that makes the integrals diverge to $-\infty$ comes from $C_n = \text{Re}\,\log F_n(i) \to -\infty$, right at the beginning of the argument, and then we only need to show that the other contributions stay bounded in $L^2(A)$. The direct implication of (c)(ii) can then be obtained by applying part (i) to $-1/F_n$.

Finally, as announced, the converse now proves itself: suppose that, in the situation of part (b), $\log F_n \to \log F$ weakly in $L^2(A)$. By compactness, $F_n \to G \in \mathcal{F}$ on a subsequence. If $G \not\equiv 0, \infty$ here, then what we already have implies that $\log F(t) = \log G(t)$ almost everywhere on A (since weak limits are unique), and this forces $G = F$. If $F_n \to 0$ or ∞ on a subsequence, then what (c) says will contradict our assumption, so this is impossible. So $G = F$ is the only possible limit point of F_n, and thus $F_n \to F$.

The same argument establishes the converse implications of part (c). □

Our next result requires some preparation before we can even formulate it. *Harmonic measure* in the upper half plane is a collection of probability measures on the boundary $\mathbb{R}$, one measure for each $z \in \mathbb{C}^+$, and it is given by

$$\omega_z(S) = \frac{1}{\pi} \int_S \frac{y}{(t-x)^2 + y^2}\, dt, \quad z = x + iy.$$

If we now have a Herglotz function F, then we can also consider $\omega_{F(z)}(S)$. This is a bounded harmonic function of $z \in \mathbb{C}^+$. Hence, if we fix S and send $y \to 0+$, $z = x + iy$, then the limit will exist for almost every $x \in \mathbb{R}$. We can thus define, for almost every $t \in \mathbb{R}$,

$$\omega_{F(t)}(S) := \lim_{y \to 0+} \omega_{F(t+iy)}(S).$$

This notation is convenient, but slightly deceptive because $\omega_{F(t)}(S)$ is not just a function of (S and) the number $F(t)$; rather, it depends on the behavior of $F(t+iy)$ as $y \to 0+$.

Let us look at this in more detail. The easiest case arises when $F(t) = \lim_{y\to 0+} F(t+iy)$ exists and lies in $\mathbb{C}^+$. Then, by dominated convergence, the limit can just be taken directly in the integral defining harmonic measure, and in this case, it is actually true that

$$\omega_{F(t)}(S) = \frac{1}{\pi} \int_S \frac{F_2(t)}{(t - F_1(t))^2 + F_2(t)^2}\, dt$$

simply is harmonic measure at the number $F(t) \in \mathbb{C}^+$.

If $F(t)$ exists, but $F(t) \in \mathbb{R}$, then

$$\omega_{F(t)}(S) = \begin{cases} 1, & F(t) \in \text{int}(S), \\ 0, & F(t) \notin \overline{S}, \end{cases}$$

as is again easy to show from the definition. So at least for nice sets S (say, intervals), $\omega_{F(t)}(S)$ essentially behaves like the Dirac measure at $F(t)$.

If $F(t) \in \mathbb{R}$ for all $t \in A$ for some $A \subseteq \mathbb{R}$ of finite measure, then this will imply that

$$\int_A \omega_{F(t)}(S)\,dt = \big|\{t \in A : F(t) \in S\}\big|, \tag{7.8}$$

so the integrals on the left-hand side give information about the distribution of $F(t)$. I mention (7.8) only to motivate what follows, so do not want to prove it in detail here. Notice that for intervals $S = (a,b)$, (7.8) is clear from what we just did because we can have $F(t) = a$ or $= b$ only on a null set. (A convenient way to establish (7.8) for arbitrary Borel sets S would be to show that both sides define measures.)

Definition 7.3. Let F_n, F be Herglotz functions. We say that $F_n \to F$ in *value distribution* if

$$\lim_{n\to\infty} \int_A \omega_{F_n(t)}(S)\,dt = \int_A \omega_{F(t)}(S)\,dt$$

for all Borel sets $A, S \subseteq \mathbb{R}$, $|A| < \infty$.

Theorem 7.5. *Let F_n be a sequence of (genuine) Herglotz functions. Then:*
(a) *if F also is a Herglotz function, then $d(F_n, F) \to 0$ if and only if $F_n \to F$ in value distribution;*
(b) $F_n \to a \in \mathbb{R}$ *if and only if*

$$\lim_{n\to\infty} \int_A \omega_{F_n(t)}(S)\,dt = \begin{cases} |A|, & a \in (c,d), \\ 0, & a \notin [c,d] \end{cases}$$

for all Borel sets $A \subseteq \mathbb{R}$, $|A| < \infty$, and all intervals $S = (c,d)$;
(c) $F_n \to \infty$ *if and only if*

$$\lim_{n\to\infty} \int_A \omega_{F_n(t)}(S)\,dt = 0$$

for all Borel sets $A \subseteq \mathbb{R}$, $|A| < \infty$, and all bounded intervals $S = (c,d)$.

The proof will follow our usual general plan (one direction will be enough), and it can be greatly simplified if we make use of the following beautiful result, which will also become very important in Section 7.5.

Theorem 7.6. *Let $A \subseteq \mathbb{R}$ be a Borel set with $|A| < \infty$. Then*

$$\lim_{y\to 0+} \sup_{F\in\mathcal{F}; S\subseteq\mathbb{R}} \left| \int_A \omega_{F(t+iy)}(S)\,dt - \int_A \omega_{F(t)}(S)\,dt \right| = 0.$$

The surprising feature of this result is that the convergence is uniform across all Herglotz functions. If $F \equiv a \in \mathbb{R}_\infty$ here, then, strictly speaking, $\omega_{F(t)}(S)$ has not been defined, but we could interpret the integrals in the same way as suggested in Theorem 7.5(b), (c) in these cases. That makes the difference equal to zero, so we can simply ignore these F here and take the supremum only over the genuine Herglotz functions.

Proof. We will make use of the elegant identity

$$\omega_{F(z)}(S) = \int_{-\infty}^{\infty} \omega_{F(t)}(S)\, d\omega_z(t), \tag{7.9}$$

which is valid for any Borel set $S \subseteq \mathbb{R}$, any $z \in \mathbb{C}^+$, and any Herglotz function F. To prove (7.9), observe that $z \mapsto \omega_{F(z)}(S)$ is harmonic, $0 \le \omega \le 1$, and $\omega_{F(t)}(S)$, by its definition, is the boundary value of this function. Thus (7.9) is the Poisson representation formula for $\omega_{F(z)}(S)$.

By Fubini's theorem,

$$\begin{aligned}\int_A \omega_{F(t+iy)}(S)\, dt &= \int_A dt \int_{-\infty}^{\infty} d\omega_{t+iy}(s)\, \omega_{F(s)}(S)\\ &= \frac{1}{\pi}\int_A dt \int_{-\infty}^{\infty} ds\, \frac{y}{(t-s)^2+y^2}\, \omega_{F(s)}(S)\\ &= \int_{-\infty}^{\infty} \omega_{F(s)}(S)\omega_{s+iy}(A)\, ds,\end{aligned}$$

and thus

$$\begin{aligned}\left|\int_A \left(\omega_{F(t+iy)}(S) - \omega_{F(t)}(S)\right) dt\right| &= \left|\int_{-\infty}^{\infty} \omega_{F(t)}(S)\left(\omega_{t+iy}(A) - \chi_A(t)\right) dt\right|\\ &\le \max_{B=A,A^c} \int_B \omega_{t+iy}(B^c)\, dt.\end{aligned}$$

The inequality follows because $0 \le \omega_{F(t)}(S) \le 1$, and the difference is negative on A and positive on A^c, so by integrating over only one of these two sets, we avoid cancellations and are making the integral larger. It is then easy to see that this final expression gives the same value for $B = A$ and $B = A^c$, so let us just take $B = A$ in the sequel.

We have estimated the difference from the statement of the theorem by

$$I(y;A) := \int_A \omega_{t+iy}(A^c)\, dt,$$

and this quantity is independent of both $F \in \mathcal{F}$ and $S \subseteq \mathbb{R}$. It remains to show that $I \to 0$ as $y \to 0+$. Almost every $t \in A$ is a point of density of A, that is, $|(t-h, t+h) \cap A^c| =$

$o(h)$ as $h \to 0+$. For such a t, we have

$$\omega_{t+iy}(A^c) \le \frac{1}{\pi} \int_{A^c \cap (t-Ny, t+Ny)} \frac{y}{(s-t)^2 + y^2}\, ds + \frac{1}{\pi} \int_{|s-t|>Ny} \frac{y}{(s-t)^2 + y^2}\, ds$$
$$= No(1) + 1 - \frac{2}{\pi} \arctan N$$

as $y \to 0+$, for any fixed $N > 0$. Since N can be arbitrarily large here, this implies that $\omega_{t+iy}(A^c) \to 0$ as $y \to 0+$ for almost every $t \in A$. Now dominated convergence shows that $I(y; A) \to 0$, as required. □

Proof of Theorem 7.5. Only the direct implications need to be shown explicitly. In the situation of part (a), we clearly have

$$\omega_{F_n(t+iy)}(S) = \frac{1}{\pi} \int_S \frac{\operatorname{Im} F_n(t+iy)}{(s - \operatorname{Re} F_n(t+iy))^2 + (\operatorname{Im} F_n(t+iy))^2}\, ds$$
$$\to \omega_{F(t+iy)}(S)$$

for fixed $y > 0$, and this implies that $F_n \to F$ in value distribution by Theorem 7.6 and dominated convergence.

If $F_n \to a \in \mathbb{R}$, then $\omega_{F_n(t+iy)}(c, d) \to \chi_{(c,d)}(a)$ for any interval $S = (c, d)$ with $c, d \ne a$, and if $F_n \to \infty$, then $\omega_{F_n(t+iy)}(c, d) \to 0$. So the argument based on Theorem 7.6 works in all cases, and we have established the direct implication of all three parts of the theorem.

The converse then follows from the compactness argument that is quite familiar by now. We also need the fact that limits in value distribution (or defined by the conditions from parts (b), (c)) are unique, so let us perhaps go through one case one more time. Suppose that $F_n \to F$ in value distribution, and F is assumed to be a (genuine) Herglotz function, corresponding to the situation of part (a). Then $F_n \to G \in \mathcal{F}$ on a subsequence, and let us look at the case when G is a Herglotz function also. In this situation,

$$\int_A \omega_{F(t)}(S)\, dt = \int_A \omega_{G(t)}(S)\, dt$$

for all Borel sets $A, S \subseteq \mathbb{R}$, $|A| < \infty$, and we want to conclude from this that $F = G$. By Lebesgue's differentiation theorem, $\omega_{F(t)}(S) = \omega_{G(t)}(S)$ for almost all $t \in \mathbb{R}$ for fixed S, and then also for any countable collection $\{S_n\}$. By throwing away another null set, we can also assume that $F(t)$, $G(t)$ both exist for these t. If we then work with the intervals with rational endpoints as our sets S_n, then it is easy to see that $F(t)$ can be reconstructed from the values of $\omega_{F(t)}(S_n)$. Thus $F(t) = G(t)$ almost everywhere, and the Herglotz functions F, G agree. The other cases, when constant functions from $\mathcal{F}$ are involved, are similar. It follows that F is the only possible limit point of the sequence F_n, and thus $d(F_n, F) \to 0$. □

7.3 Reflectionless canonical systems

Definition 7.4. For $H \in \mathcal{C}(\mathbb{R})$, we define the *reflection coefficients* as

$$R_+(z) = \frac{\overline{m_+(z)} + m_-(z)}{m_+(z) + m_-(z)}, \quad R_-(z) = \frac{m_+(z) + \overline{m_-(z)}}{m_+(z) + m_-(z)}$$

if $H \not\equiv P_\alpha$, and if $H \equiv P_\alpha$, then we put $R_\pm(z) = 0$.

In the first case, it could happen that m_+ or $m_- \equiv \infty$ (but not both, this would put us into the second case), and when it does, then our formulae do not really work very well, not even in a limiting sense; we simply set $R_+ = R_- = 1$ in this case. The reflection coefficients are then well defined for $z \in \mathbb{C}^+$ in all cases. They satisfy $|R_\pm(z)| \le 1$, and, since $m_+ + m_- \not\equiv 0$ unless $H \equiv P_\alpha$, they have boundary values $R_\pm(x) = \lim_{y\to 0+} R_\pm(x+iy)$ at almost all $x \in \mathbb{R}$.

We will be mostly interested in their absolute values, and clearly $|R_+(z)| = |R_-(z)|$, so I will usually just write $|R(z)|$ then.

For the classical equations (Schrödinger, Jacobi, Dirac), these quantities are indeed related to the reflection coefficients of stationary scattering theory. This will not play a role here, though, and in any event, this theory does not generalize in a natural way due to the lack of a comparison operator for canonical systems.

The reflection coefficients allow us to recover the absolutely continuous spectrum, up to unitary equivalence. Recall that this is really the same as finding the essential supports

$$\Sigma_{ac}^\pm = \{t \in \mathbb{R} : \operatorname{Im} m_\pm(t) > 0\},$$

as discussed in Section 3.7. Up to sets of Lebesgue measure zero, these sets can also be described in terms of the reflection coefficients as

$$\Sigma_{ac}^\pm = \{t \in \mathbb{R} : R_\pm(t) \ne 1\}.$$

Here we must exclude the trivial canonical systems $H \equiv P_\alpha$; this is no big problem since $D(\mathcal{S}) = 0$ in this case and thus there is no spectral theory that could be discussed. Somewhat annoyingly however, we also have to exclude the canonical systems with m_- or $m_+ \equiv \infty$, since we simply set $R_\pm = 1$ in this case, so the reflection coefficients no longer carry any information about the other m function.

Fortunately, the reflection coefficients do give us a very convenient description of the intersection of these half line supports as

$$\Sigma_{ac} = \{t \in \mathbb{R} : |R(t)| < 1\}, \tag{7.10}$$

and this now works in all cases as long as $H \not\equiv P_\alpha$. Recall also that Σ_{ac} was the set where the (local) multiplicity equals two.

The absolutely continuous spectra themselves can then be obtained as the essential closures of these sets, and the *essential closure* is defined as the closure, but ignoring null sets: by definition, an $x \in \mathbb{R}$ is in the essential closure of $A \subseteq \mathbb{R}$ if and only if $A \cap (x-h, x+h)$ has positive measure for all $h > 0$.

Toda maps not only preserve all spectral properties; they also preserve the absolute values of the reflection coefficients.

Theorem 7.7. *Suppose that $H_j \in \mathcal{C}(\mathbb{R})$ are related by a Toda map. Then $|R_2(x)| = |R_1(x)|$ for almost every $x \in \mathbb{R}$.*

Proof. First of all, if H_1 or $H_2 \equiv P_\alpha$, then both canonical systems will be of this type, and $R_1 = R_2 = 0$. Also, if one reflection coefficient, say R_1, satisfies $|R_1(x)| = 1$ on a set $A \subseteq \mathbb{R}$ of positive measure, then we will also have $|R_2(x)| = 1$ almost everywhere on A. This follows from Theorem 7.2: if we had $|R_2| < 1$ on a set $B \subseteq A$, $|B| > 0$, then H_2 would have absolutely continuous spectrum of multiplicity two essentially supported by (at least) B, while H_1 does not have this property, so the operators could not be unitarily equivalent in this scenario.

So we can focus on the (common) set where $|R_j(x)| < 1$. By throwing away null sets if needed, we may also assume that $m_\pm^{(j)}(x)$, $j = 1, 2$, exist on this set and lie in $\mathbb{C}^+$.

Now in principle the proper way to obtain the boundary values $m_\pm^{(2)}(x)$ of the transformed functions would be to apply $S(z)$ to $\pm m_\pm^{(1)}(z)$ for $z = x + iy$, $y > 0$, and then send $y \to 0+$. However, since S is entire, $S(x) \in SL(2, \mathbb{R})$, and we are currently considering $x \in \mathbb{R}$ for which $m_\pm^{(1)}(x) \in \mathbb{C}^+$, we can also perform these operations in the reverse order and simply apply $S(x)$ directly to $\pm m_\pm^{(1)}(x)$. The same remarks apply to $R(x)$: we can first take the boundary values of the m functions and then plug these into the formula defining R, and this will give the correct answer at almost all $x \in \mathbb{R}$.

Now a quick calculation will show that indeed $|R(x)|$ is preserved under such a transformation by $S(x) \in SL(2, \mathbb{R})$. □

The fact stated in the last sentence of this proof could be given a more abstract explanation. Namely, if $m_\pm(z) \in \mathbb{C}^+$, then $|R(z)|$ essentially computes the hyperbolic distance of m_+ and $-\overline{m_-}$, which is defined, for two points $w, z \in \mathbb{C}^+$, as

$$\delta(w, z) = \tanh^{-1}\left|\frac{z-w}{z-\overline{w}}\right| \tag{7.11}$$

(usually an additional factor of 2 is introduced on the right-hand side, but this will not matter here), and thus

$$|R(z)| = \tanh \delta\left(m_+(z), -\overline{m_-(z)}\right).$$

Holomorphic self-maps of $\mathbb{C}^+$ (in other words, Herglotz functions) decrease hyperbolic distance, and thus the action of an $S \in SL(2, \mathbb{R})$, being an automorphism, preserves it.

Definition 7.5. Let $A \subseteq \mathbb{R}$ be a Borel set. We call a canonical system $H \in \mathcal{C}(\mathbb{R})$ *reflectionless* on A, and we write $H \in \mathcal{R}(A)$, if $R(x) = 0$ for almost every $x \in A$.

Note that the trivial canonical systems $H \equiv P_\alpha$ are reflectionless on $A = \mathbb{R}$. This may seem strange at first; for example, reflectionless operators usually have absolutely continuous spectrum on the set on which they are reflectionless; see Theorem 7.10 below. However, as we will see later, the key results of the next two sections simply will not work if these canonical systems consisting of a single singular interval are not declared to be reflectionless everywhere. In fact, this is precisely the reason why I set $R = 0$ for these systems in Definition 7.4.

As we will see in more detail soon, reflectionless canonical systems are quite rare. They are important because they form the basic building blocks of arbitrary canonical systems with non-empty absolutely continuous spectrum. Let us collect some of their basic properties.

Proposition 7.8. *Let $H \in \mathcal{C}(\mathbb{R})$, $H \not\equiv P_\alpha$, and let $A \subseteq \mathbb{R}$ be a Borel set. The following statements are equivalent:*
(a) $H \in \mathcal{R}(A)$;
(b) $m_+(x) = -\overline{m_-(x)}$ *for almost every* $x \in A$;
(c) $\operatorname{Re} M(x) = 0$ *for almost every* $x \in A$.

Proof. The equivalence of (a) and (b) is obvious from the definition of $\mathcal{R}(A)$. A straightforward calculation shows that (b) implies (c). Conversely, if (c) holds, then we can focus on the two diagonal entries of $\operatorname{Re} M(x)$, and from this, we obtain

$$\operatorname{Re}\left(m_+(x) + m_-(x)\right) = 0,$$
$$|m_+(x)|^2 \operatorname{Re} m_-(x) + |m_-(x)|^2 \operatorname{Re} m_+(x) = 0,$$

and this quickly implies (b). It is also worth observing here that (c) cannot hold if m_+ or $m_- \equiv \infty$, so we can carry out these calculations without having to worry about this scenario. □

Theorem 7.9. (a) *Suppose that $H_j \in \mathcal{C}(\mathbb{R})$, $j = 1, 2$, are related by a Toda map. Then $H_1 \in \mathcal{R}(A)$ if and only if $H_2 \in \mathcal{R}(A)$.*
(b) *Suppose that $|A| > 0$. If $H_j(x) \in \mathcal{R}(A)$, $j = 1, 2$, agree on a half line $x \in (L, \infty)$ (or $x \in (-\infty, L)$), then $H_1 = H_2$.*

Part (b) will later lead to the property of general canonical systems with non-empty absolutely continuous spectrum of having values that can be approximately predicted (for a while at least) from past values. It also makes more precise my earlier remark that reflectionless systems are very rare.

Proof. (a) This is immediate from Theorem 7.7.

(b) By part (a), since we can apply (plain) shifts to canonical systems without leaving $\mathcal{R}(A)$, we can assume that $L = 0$. Then, if $H_1 = H_2$ on, say, $(0, \infty)$, then $m_+^{(1)}(z) = m_+^{(2)}(z)$, and then Proposition 7.8(b) shows that also $m_-^{(1)}(x) = m_-^{(2)}(x)$ for almost every $x \in A$. Since $|A| > 0$, these boundary values determine the Herglotz functions $m_-^{(j)}(z)$ completely, so $m_-^{(1)}(z) = m_-^{(2)}(z)$ for arbitrary $z \in \mathbb{C}^+$.

This argument also works if $m_+^{(j)} \equiv a \in \mathbb{R}_\infty$ because it is then still true that the only way to obtain a reflectionless canonical system is to make $m_- \equiv -a$; this follows from Proposition 7.8 or from the definition of reflectionless systems. □

Theorem 7.10. *Let $H \in \mathcal{R}(A)$, $H \not\equiv P_\alpha$. Then $\Sigma_{ac} \supseteq A$.*

Proof. This is obvious from (7.10). □

Theorem 7.11. *The set $\mathcal{R}(A)$ is a compact subset of $(\mathcal{C}(\mathbb{R}), d)$.*

This will also follow from Theorem 7.12 of the next section but since it has a direct proof that is much easier than what we will do there, it seems appropriate to include it here.

Proof. The set $\mathcal{C}(\mathbb{R})$ is compact, so we must show that $\mathcal{R}(A) \subseteq \mathcal{C}(\mathbb{R})$ is closed. Since the singular lines $H \equiv P_\alpha$, $0 \le \alpha < \pi$, obviously form a closed subset, we can ignore them here and assume that we have a convergent sequence $H_n \to H$, $H_n \in \mathcal{R}(A)$, where the H_n are not of this type.

By Proposition 7.8 and its proof, it suffices to show that the two Herglotz functions

$$F(z) = m_+(z) + m_-(z), \quad G(z) = -\frac{1}{m_+(z)} - \frac{1}{m_-(z)} \tag{7.12}$$

have real part zero almost everywhere on A; note that these functions are the negative reciprocals of the diagonal entries of M. Of course, these formulae cannot be taken at face value if at least one of the m functions is identically equal to zero or infinity, so let us first deal with this special scenario. Let us say $m_- \equiv \infty$. We have $m_-(z; H_n) \to m_-(z; H)$ and $m_+(x; H_n) = -\overline{m_-(x; H_n)}$ almost everywhere on A, so Theorem 7.4(c) now shows that $m_+(z; H_n) \to \infty$ as well, so $H \equiv P_{e_2}$, and we have the desired conclusion that $H \in \mathcal{R}(A)$. The other (degenerate) cases are similar.

In the remaining case, when neither m_+ nor m_- is identically equal to zero or infinity, we return to (7.12) and look at the corresponding functions F_n, G_n, formed in the same way, but with $m_\pm(z; H_n)$. By assumption, we have $\operatorname{Re} F_n(x) = 0$ on A, so $\operatorname{Im} \log F_n(x) = \pi/2$, and this property is preserved when we pass to the weak limit, as in Theorem 7.4(b). It could again happen here that we are really in case (c) of that theorem, when $m_+ + m_- \equiv 0$, but, as above, this gives $H \equiv P_\alpha$. The discussion of G is of course analogous, and we conclude that $H \in \mathcal{R}(A)$ in all cases. □

7.4 Semi-continuity of R and Σ_{ac}

Theorem 7.12. *Let $H_n, H \in \mathcal{C}(\mathbb{R})$ and suppose that $H_n \to H$. Then*

$$|R(x; H)| \le \limsup_{n \to \infty} |R(x; H_n)|$$

for almost all $x \in \mathbb{R}$.

Of course, this makes a non-trivial statement only at those x at which $\limsup |R(x;H_n)| < 1$. A particularly interesting situation arises when we produce the whole sequence by applying Toda maps to a fixed $H_1 \in \mathcal{C}(\mathbb{R})$. Then $|R(x;H_n)|$ is constant, and now the theorem says that $|R(x;H)|$ can only go down, after passing to the limit. By making use of (7.10), we then obtain from this a corresponding semi-continuity property of the essential supports Σ_{ac} of the absolutely continuous parts of multiplicity two.

Corollary 7.13. *Suppose that each H_n is related to H_1 by a Toda map, and $H_n \to H$. Then $|R(x;H)| \le |R(x;H_1)|$ for almost all $x \in \mathbb{R}$, and if $H \not\equiv P_\alpha$, then $\Sigma_{ac}(H) \supseteq \Sigma_{ac}(H_1)$.*

Even though this is a special case of the theorem, we still have a large supply of transformations at our disposal that can be applied to H_1. However, as the corollary shows, only special limits can ever be reached if we start out with non-empty absolutely continuous spectrum of multiplicity two. In the next section, we will show that if the limits are taken under twisted shift maps, then only reflectionless canonical systems can be limit points, and this is an extremely restrictive condition.

Proof of Theorem 7.12. We must show that if $C < 1$ and $L > 0$ are given and we define

$$B = \left\{ x \in (-L, L) : \limsup_{n\to\infty} |R(x;H_n)| < C \right\},$$

then $|R(x;H)| \le C$ for almost every $x \in B$. Note that we do not need the strict inequality in this statement even though this is what the theorem asserts; it will follow automatically for almost all x since we can vary C. We can go further here by considering subsets $A \subseteq B$ of the type

$$A = \{x \in (-L, L) : |R(x;H_n)| \le C \text{ for all } n \ge N\}.$$

Since $|B \setminus A| \to 0$ as $N \to \infty$, it actually suffices to show that $|R(x)| \le C$ for almost all $x \in A$, for any such A.

As a second preparation of a technical nature, observe that we can transform all canonical systems H_n, H (or rather their m functions) by a fixed automorphism $S \in SL(2,\mathbb{R})$, and since this does not change $|R_n|$ or $|R|$, it will not affect the statement. We may and will thus assume that $m_\pm(z;H) \not\equiv \infty$. Moreover, the case when $m_+(z;H) + m_-(z;H) \equiv 0$ is trivial because then $H \equiv P_\alpha$, so we can also assume that we are not in this situation. As a consequence, we then also have $m_+(z;H_n) + m_-(z;H_n) \not\equiv 0$ for all large n.

Now consider the functions $L_\pm(x;H_n) = \log(R_\pm(x;H_n) - 1)$, $x \in A$. Here and in the sequel, we fix the choice of the logarithm by agreeing that $\text{Im}\,\log w \in [0, 2\pi)$. The sequence $L_\pm(x;H_n)$ is bounded in $L^2(A)$, so converges weakly on a subsequence, which we will as usual not make explicit in the notation. Moreover, we have

$$R_\pm - 1 = -\frac{2i\text{Im}\, m_\pm}{m_+ + m_-},$$

so

$$L_\pm(x;H_n) = \log(-2i) - \log F(x;H_n) + \log \operatorname{Im} m_\pm(x;H_n), \tag{7.13}$$

and here I have abbreviated $F = m_+ + m_-$. Now we apply Theorem 7.4 to the sequence $\log F(x;H_n)$. We know that $F(z;H_n) \to F(z;H)$, and this latter function is not identically equal to zero or infinity, by the extra assumptions we made above. So we are then in case (b) of Theorem 7.4 and $\log F(x;H_n) \to \log F(x;H) \in L^2(A)$. Now (7.13) shows that

$$\log \operatorname{Im} m_\pm(x;H_n) \to P_\pm(x) \in L^2(A)$$

weakly in $L^2(A)$, for some functions $P_\pm$, and the convergence really takes place on the subsequence that was chosen earlier. I now claim that this gives us the following estimate, for almost every $x \in A$:

$$\operatorname{Im} m_\pm(x;H) \ge e^{P_\pm(x)}. \tag{7.14}$$

To prove this, drop the reference to the sign $\pm$ for now, to simplify the notation, and let $I = (a,b)$ be a bounded interval with $\rho(\{a\}) = \rho(\{b\}) = 0$, and here ρ is the measure associated with m; the measures of $m_n(z) = m(z;H_n)$ will similarly be denoted by ρ_n. Then

$$\begin{aligned}
\frac{\pi\rho(I)}{|I|} &= \lim_{n\to\infty} \frac{\pi\rho_n(I)}{|I|} \ge \limsup_{n\to\infty} \frac{1}{|I|} \int_{I\cap A} \operatorname{Im} m_n(x)\, dx \\
&\ge \frac{|I\cap A|}{|I|} \limsup_{n\to\infty} \exp\left(\frac{1}{|I\cap A|} \int_{I\cap A} \log \operatorname{Im} m_n(x)\, dx \right) \\
&= \frac{|I\cap A|}{|I|} \exp\left(\frac{1}{|I\cap A|} \int_{I\cap A} P(x)\, dx \right);
\end{aligned}$$

the second estimate is Jensen's inequality. Now take $I = (x-h, x+h)$, and send $h \to 0+$ through a sequence chosen as above, such that ρ does not give weight to the endpoints. At almost every $x \in A$, the following statements will be true:

$$\frac{|I\cap A|}{|I|} \to 1, \quad \frac{1}{|I|} \int_{I\cap A} P(t)\, dt \to P(x), \quad \frac{\pi\rho(I)}{|I|} \to \operatorname{Im} m(x).$$

Thus we obtain (7.14) at all such $x \in A$.

In particular, in the case currently under consideration, when $H \not\equiv P_\alpha$, we then obtain $\operatorname{Im} m_\pm(x;H) > 0$ almost everywhere on A. (Note that this already proves the statement of Corollary 7.13 about $\Sigma_{ac}(H)$, so if we are only interested in this and not the reflection coefficients themselves, then we could stop here.)

To prove that the reflection coefficients are semi-continuous, too, we return to (7.13) and its limiting version, as $n \to \infty$. We introduce two new functions $r_\pm(x)$ by writing

$$\lim_{n\to\infty} \log(R_\pm(x;H_n) - 1) = \log(r_\pm(x) - 1);$$

more precisely, we define $r_\pm$ by this and the additional requirement that $|r_\pm(x)| \le C$. This is possible because the values of a weak limit in $L^2(A)$ are contained in the closed convex hull of the values of the approximating functions, and we then have the following elementary fact available, whose proof we postpone.

Lemma 7.14. *The set* $\{\log(z-1) : |z| \le C\}$ *is a convex subset of* $\mathbb{C}$ *for any* $C < 1$.

After passing to the limit, (7.13) then gives the identity

$$\log(R_\pm - 1) = \log(r_\pm - 1) + \log \operatorname{Im} m_\pm - P_\pm; \tag{7.15}$$

all functions refer to the canonical system H, and they are evaluated at $x \in A$. Note also that the weak limit of $\log(R(x;H_n) - 1)$ need not be equal to $\log(R(x;H) - 1)$; this would have been an unreasonably optimistic expectation.

Take the real and imaginary parts of (7.15) and use (7.14) to estimate the real part. This yields

$$\operatorname{Re}\, \log(R_\pm - 1) \ge \operatorname{Re}\, \log(r_\pm - 1), \tag{7.16}$$

$$\operatorname{Im}\, \log(R_\pm - 1) = \operatorname{Im}\, \log(r_\pm - 1) = \frac{3\pi}{2} - \operatorname{Im}\, \log F; \tag{7.17}$$

the final equality follows from (7.13).

We will now obtain the desired conclusion that $|R_\pm| \le C$ from an elementary analysis of (7.16), (7.17) at a fixed $x \in A$. Write $R_\pm - 1 = \rho_\pm e^{i\varphi}$, with $\pi/2 < \varphi < 3\pi/2$; both numbers indeed have the same argument, by (7.17). We can now rephrase (7.16) as follows: define ρ_0 as the absolute value of the smallest point of intersection of the ray $te^{i\varphi}$, $t \ge 0$, with the circle $-1 + Ce^{i\alpha}$; in other words, since $|r_\pm| \le C$ and $r_\pm - 1$ also has argument φ, this is the smallest value that $|r_\pm - 1|$ can possibly take. Then $\rho_\pm \ge \rho_0$. Moreover, the inequality $|R_\pm| \le C$ that we want to establish is now equivalent to the claim that $\rho_\pm \le \rho_1$, with ρ_1 defined as the absolute value of the larger point of intersection of the ray with the circle.

By elementary geometry in the plane, we find

$$\rho_j = -\cos\varphi \pm \sqrt{C^2 - \sin^2\varphi}, \quad j = 0, 1; \tag{7.18}$$

more precisely, taking the plus sign gives us ρ_1. Recall also here that $\pi/2 < \varphi < 3\pi/2$, so $\cos\varphi < 0$. Moreover, the ray and the circle must intersect or we would obtain a contradiction to (7.17), and this will make sure that $|\sin\varphi| \le C$.

Now consider

$$-\frac{\overline{R_+}}{R_-} = e^{i\pi}\frac{F}{\overline{F}} = e^{-2i\varphi};$$

the second equality follows from (7.17). So

$$R_- = -\overline{R_+}e^{2i\varphi} = -(1 + \rho_+ e^{-i\varphi})e^{2i\varphi},$$

and hence $R_- - 1 = (-\rho_+ - 2\cos\varphi)e^{i\varphi}$, and this yields

$$\rho_- = -2\cos\varphi - \rho_+ \equiv f(\rho_+).$$

From (7.18) we see that $f(\rho_1) = \rho_0$, and since f is a strictly decreasing function of ρ_+, this means that if we had $\rho_+ > \rho_1$, then it would follow that $\rho_- < \rho_0$, which contradicts the inequality $\rho_\pm \geq \rho_0$ that we observed above. We have shown that $\rho_+ \leq \rho_1$, and, as discussed, this says that $|R_+| \leq C$. □

Proof of Lemma 7.14. The set under consideration can be described as

$$\{x + iy : 1 - C \leq e^x \leq 1 + C, \varphi(x) \leq y \leq 2\pi - \varphi(x)\},$$

and here $\varphi(x) \in (\pi/2, \pi]$ is defined as the angle for which $e^{x+i\varphi(x)}$ lies on the circle $\{-1 + z : |z| = C\}$. It will now be enough to show that the complementary angle $\psi(x) = \pi - \varphi(x)$ is a concave function of x in the range indicated.

Elementary geometry gives us the formula

$$\cos\psi(x) = \frac{1}{2}(e^x + (1 - C^2)e^{-x}),$$

so $(\cos\psi)'' = \cos\psi$, and when written out, this becomes

$$-\psi'' \sin\psi = (1 + \psi'^2)\cos\psi.$$

Since $0 \leq \psi < \pi/2$, this implies that $\psi'' \leq 0$, as required. □

7.5 Reflectionless limit points

We continue the investigations of the previous section. We will now take limits along a sequence of twisted shifts of a given canonical system.

Definition 7.6. Let $H \in \mathcal{C}(\mathbb{R})$. The *$\omega$ limit set* $\omega(H)$ of H is defined as the collection of all limit points $K = \lim H_n$, where H_n is a twisted shift of H of length x_n, and $x_n \to \infty$.

Equivalently, we can say that $K \in \omega(H)$ if and only if

$$\pm m_\pm(z; K) = \lim_{n\to\infty} A_n T(x_n; z; H)(\pm m_\pm(z; H))$$

for some $A_n \in SL(2, \mathbb{R})$ and some sequence $x_n \to \infty$.

In the sequel, I will sometimes write a twisted shift of H of length x as $x \cdot H$. This notation is not very illuminating since it suppresses the dependence of the twisted shift on $A \in SL(2, \mathbb{R})$, but it will be convenient anyway as a compact short-hand notation.

When we form the ω limit set $\omega(H)$, we in particular collect all limit points under plain shifts. As discussed in Section 7.1, the class of twisted shift maps is a more natural and useful choice for canonical systems than just considering the shifts themselves.

Observe that always $\omega(H) \neq \emptyset$ by the compactness of $\mathcal{C}(\mathbb{R})$, and $\omega(H)$ is itself compact; another easy consequence of compactness is the following, which will be used below.

Proposition 7.15. *Let $S_x(H) = \{y \cdot H : y \geq x, A \in \mathrm{SL}(2,\mathbb{R})\}$ be the collection of all twisted shifts of a given $H \in \mathcal{C}(\mathbb{R})$ of length at least x, and define $d(x) = \sup_{K \in S_x(H)} d(K, \omega(H))$. Then $d(x) \to 0$ as $x \to \infty$.*

Proof. If this were false, then there would be a $\delta > 0$ such that $d(x_n \cdot H, \omega(H)) \geq \delta$ for certain twisted shifts of H of lengths $x_n \to \infty$. Obviously, this is impossible, since this sequence has limits on subsequences, which must lie in $\omega(H)$. □

The ω limit set is not just non-empty, it is also clear that it always contains all $K \equiv P_\alpha$. To see this, observe that $\omega(H)$ is invariant under the action of $\mathrm{SL}(2,\mathbb{R})$. Moreover, for any given sequence $x_n \to \infty$, we can take

$$A_n = \begin{pmatrix} c_n & 0 \\ 0 & 1/c_n \end{pmatrix};$$

this matrix acts by multiplication by c_n^2, and if the c_n now converge to zero sufficiently rapidly, then we would expect that $m_\pm(z; x_n \cdot H) \to 0$, so that $K \equiv P_{e_1}$ then lies in $\omega(H)$. More precisely, the only way this can fail is if either m_- or $m_+(z; x_n \cdot H) \equiv \infty$ for infinitely many n, but it is easy to see that $K \equiv P_{e_2} \in \omega(H)$ in that case.

Let us also state these easy properties as a formal result.

Proposition 7.16. *For any $H \in \mathcal{C}(\mathbb{R})$, if $K \in \omega(H)$, then the trace normed version of $A^t K(x) A$ will be in $\omega(H)$ also, for any $A \in \mathrm{SL}(2,\mathbb{R})$. Moreover,*

$$\omega(H) \supseteq \{K \equiv P_\alpha : 0 \leq \alpha < \pi\}.$$

It is natural to expect $\omega(H)$ to be invariant under general twisted shifts. Furthermore, this set should only depend on the restriction of H to any right half line, for example to $(0,\infty)$. Both properties can look completely obvious if one is not careful, but actually the unqualified versions are false in both cases: we must not forget that we need the *trace normed* version of

$$(L \cdot H)(x) = A^{-1t} H(x + L) A^{-1}, \tag{7.19}$$

and when we do normalize the trace, the lengths of intervals can change dramatically when passing to the new variable. For example, suppose that H is known on $(0,\infty)$, and we want to find $\omega(H)$ from this information. Then the interval $(-L, 0)$, on which the restriction to $(-\infty, 0)$ of the function from (7.19) is known, could shrink considerably. More precisely, its length becomes

$$\mathrm{tr} \int_0^L A^{-1t} H(x) A^{-1}\, dx$$

after the normalization. So matters are clarified by the following observation.

Lemma 7.17. *Suppose that $H \in \mathcal{C}$, $H \not\equiv P_\alpha$. Then*

$$\inf_{B\in SL(2,\mathbb{R})} \operatorname{tr} \int_0^L B^t H(x) B\, dx \to \infty$$

as $L \to \infty$.

Note that here H is assumed to be a half line canonical system, so if we want to apply this to an $H \in \mathcal{C}(\mathbb{R})$, then what we need to assume is that $H \not\equiv P_\alpha$ on $(0, \infty)$.

As discussed, the lemma implies that if $(0, \infty)$ is not contained in a single singular interval, then $\omega(H)$ can be found from the right half line restriction of H. There are similar consequences regarding the invariance of $\omega(H)$ under twisted shifts that I mentioned briefly above; we do not need these here, so I will not discuss them further.

Proof. Let $M = M(L) = \int_0^L H(x)\,dx$. These matrices are positive definite and increasing, in the sense that $M(L_2) - M(L_1)$ also is a positive definite matrix if $L_2 \ge L_1$. Suppose now that $\operatorname{tr} B^t M(L) B$ stays bounded as $L \to \infty$ for suitably chosen matrices $B = B(L) \in SL(2, \mathbb{R})$. Since $\operatorname{tr} M(L) = L$, at least one diagonal entry of this matrix will be $\ge L/2$, and let us say $M_{11} \ge L/2$. We must then have $\|B^{-1}e_1\| \gtrsim L^{1/2}$. Since $\det |B^{-1}| = \det B^{-1} = 1$, this implies that there also is a unit vector $v = v(L) \in \mathbb{R}^2$ such that $\|B^{-1}v\| \lesssim L^{-1/2}$. By testing on the vector $B^{-1}v$, we then find that $v^* M v \lesssim 1/L$. So $M(L)$ has an eigenvalue $\lesssim 1/L$. Since L is arbitrary here and $M(L)$ is increasing, this implies that $M(L)$ has zero as an eigenvalue for all $L > 0$. Clearly, this is only possible if $H \equiv P_\alpha$. □

Now that this has been settled, we could also assume in the following result that we are given a half line canonical system $H \in \mathcal{C}$, and in fact that would perhaps be the more natural formulation.

Theorem 7.18. *Let $H \in \mathcal{C}(\mathbb{R})$ and denote the essential support of the absolutely continuous part of the spectral measure of the system on $(0, \infty)$ by*

$$\Sigma_{ac}^+ = \{t \in \mathbb{R} : \operatorname{Im} m_+(t) > 0\}.$$

Then $\omega(H) \subseteq \mathcal{R}(\Sigma_{ac}^+)$.

Compare this with Theorem 7.12, or, better yet perhaps, with Corollary 7.13: there we showed that the reflection coefficients of any limit point cannot be larger than those of the approximating sequence, and of course nothing more can be said in this generality since for example the sequence could be constant. Theorem 7.18, on the other hand, shows that the reflection coefficients drop all the way down to zero if we specifically act by twisted shifts of arbitrarily large lengths. This builds an irreversibility into the evolution, and we obtain a much stronger statement in this situation.

As I already pointed out several times, not many canonical systems are reflectionless, so Theorem 7.18 restricts the asymptotics of a canonical system with non-empty absolutely continuous spectrum quite severely. This effect becomes particularly pronounced when the set Σ_{ac}^+ has a simple topological structure. For a quick illustration, suppose that we have a half line canonical system whose spectral measure contains an absolutely continuous part equivalent to Lebesgue measure on $\mathbb{R}$. In other words, $\Sigma_{ac}^+ = \mathbb{R}$ (and then also $\sigma_{ac} = \mathbb{R}$, but this latter statement is strictly weaker). What are the possible ω limit points for such a canonical system H?

As always, $\omega(H)$ will contain all $K \equiv P_\alpha$. If $K \in \omega(H)$ is not of this type, then $F(z;K) = m_+ + m_- \not\equiv 0$, and this function has zero real part almost everywhere on $\mathbb{R}$. This property makes its Krein function identically equal to $\xi = 1/2$. A Herglotz function can be reconstructed from its Krein function, up to a positive multiplicative constant, and thus $F(z) = 2Ci$, for some $C > 0$. This could be derived systematically, but it is much easier to just observe that this F has the correct Krein function. Since $F = m_+ + m_-$ and $\text{Im}\, m_+ = \text{Im}\, m_-$ almost everywhere on $\mathbb{R}$, this implies that $\text{Im}\, m_\pm(x) = C$. Since the measures $\rho_\pm$ of $m_\pm$ are absolutely continuous with respect to the measure of F, which is $2C/\pi$ times Lebesgue measure, it follows that we already have those measures $\rho_\pm$, and they satisfy $\pi d\rho_\pm = C\, dt$. This determines $m_\pm$, up to an additive real constant, and $\text{Re}\, m_+ = -\text{Re}\, m_-$ on $\mathbb{R}$. We have found that if $K \in \mathcal{R}(\mathbb{R})$ is not a singular line, then

$$m_\pm(z;K) = \pm a + iC \tag{7.20}$$

for some $C > 0$, $a \in \mathbb{R}$. Conversely, these m functions obviously give us canonical systems that are reflectionless on $\mathbb{R}$. Let us summarize and rephrase slightly.

Theorem 7.19. *Let $H \in \mathcal{C}(\mathbb{R})$. Then $H \in \mathcal{R}(\mathbb{R})$ if and only if $H(x)$ is constant on $\mathbb{R}$.*

Proof. We found that if $H \in \mathcal{R}(\mathbb{R})$, then $H \equiv P_\alpha$, which obviously is a constant function of $x \in \mathbb{R}$, or the m functions are given by (7.20). It is easy to show that specifically $m_\pm(z;H) = i$ are the m functions of $H(x) = 1/2$. This could be done by a straightforward calculation, taking advantage of the fact that this canonical system can be solved explicitly. Alternatively, it also follows at once from Theorems 6.13, 6.15. General constants a, C in (7.20) can then be obtained from these m functions by acting by a suitable $A \in SL(2,\mathbb{R})$, and this will give us the coefficient functions $A^{-1t}(1/2)A^{-1}$, which are obviously also constant (and which would have to be normalized to make the trace equal to one again).

Since any positive definite matrix of trace one either equals P_α, for some α, or is a positive multiple of B^tB, for some $B \in SL(2,\mathbb{R})$, it is clear that, conversely, we do obtain all constant coefficient functions here. □

Before we return to the general theory, let me give another concrete example that will actually make an important general point. As we observed above, in Proposition 7.16, we always have the canonical systems $K \equiv P_\alpha$ in the ω limit set $\omega(H)$ of any $H \in \mathcal{C}(\mathbb{R})$, and the reason for this was quite trivial: since we can act on the m functions by any $SL(2,\mathbb{R})$ matrix, we can in particular implement multiplication by a small or large number, and this will make $m_\pm$ converge to 0 or ∞, respectively. From this perspective, the canonical systems $K \equiv P_\alpha$ might appear to be an artificial addition to the ω limit set that does not really tell us much about the asymptotics of $H(x)$ as $x \to \infty$.

It is important to realize, however, that we often obtain these limit points also under plain shifts. For a rather natural example where this happens, consider the Schrödinger equation with zero potential

$$-y'' = zy$$

on $x \in \mathbb{R}$. This has purely absolutely continuous spectrum of multiplicity two on $\Sigma_{ac} = [0, \infty)$. Now write this as a canonical system, using the method from Section 1.3. We need the solutions p, q of $-y'' = 0$ with the initial values $p(0) = q'(0) = 0$, $p'(0) = q(0) = 1$, and then $H_0 = \begin{pmatrix} p^2 & pq \\ pq & q^2 \end{pmatrix}$. Obviously, these are given by $p(x) = x$, $q(x) = 1$, so

$$H_0(x) = \begin{pmatrix} x^2 & x \\ x & 1 \end{pmatrix}.$$

To pass to a trace normed version H of this coefficient function, we must introduce the new variable $X = \int_0^x \operatorname{tr} H_0(t)\, dt = x + x^3/3$ and then form $H(X) = H_0/(1+x^2)$. For large X, this will satisfy $H(X) = P_{e_1} + O(X^{-1/3})$, and thus $H(X + L) \to P_{e_1}$, when we apply plain shifts. So we not only obtain this coefficient function $K \equiv P_{e_1}$ as a limit point without the need to act by $SL(2, \mathbb{R})$ matrices, but it is in fact the only limit point under plain shifts.

Let us now return to the general theory. One consequence of Theorem 7.18 is the fact that if absolutely continuous spectrum is present, then the coefficient function can be approximately predicted, for a while at least. I already mentioned this briefly above. Let us now make it precise. This will involve restrictions of coefficient functions H to the half lines $(-\infty, 0)$ and $(0, \infty)$, and I will use self-explanatory subscript notation $H_\pm$, $\mathcal{C}_\pm$, etc. for that.

Theorem 7.20 (Oracle theorem). *Let a Borel set $A \subseteq \mathbb{R}$, $|A| > 0$, and $\epsilon > 0$ be given. Then there is a function $\Delta : \mathcal{C}_- \to \mathcal{C}_+$ (the oracle) with the following property: if $H \in \mathcal{C}(\mathbb{R})$ is any canonical system with $\Sigma_{ac}^+(H) \supseteq A$, then there is an $x_0 > 0$ such that if we take any twisted shift of H of length $x \geq x_0$, then*

$$d_+(\Delta((x \cdot H)_-), (x \cdot H)_+) < \epsilon.$$

I have formulated this for arbitrary twisted shifts and it works in this generality, but we can most easily interpret it if we focus on plain shifts for the moment. Then the oracle will predict the right half line (x, ∞) approximately, based on information about past values, on $(-\infty, x)$. However, since the prediction is only valid up to a small error ϵ and the metric does not care much about what happens very far out, what is really going on is that H on $(x - L, x)$ for a sufficiently large $L > 0$ will approximately determine H on $(x, x + L')$, and here the L' that can be achieved will depend on what value for ϵ was chosen originally.

Let us now prove Theorem 7.20, assuming Theorem 7.18; this latter result will then be proved next.

Proof. Given $\epsilon > 0$, find a $\delta > 0$ such that $d_+(K_+, K'_+) < \epsilon$ for any $K, K' \in \mathcal{R}(A)$ with $d_-(K_-, K'_-) < 2\delta$. This is possible because the restriction map $K \mapsto K_-$ is obviously continuous, and it is injective on $\mathcal{R}(A)$ by Theorem 7.9(b). It then also maps between compact metric spaces, by Theorem 7.11, and thus its inverse $K_- \mapsto K$ is uniformly continuous. If we then compose this map with the restriction to the right half line, then the resulting map $K_- \mapsto K_+$ is still uniformly continuous, and this is what we used above. We can also demand here that $\delta < \epsilon$.

Now consider the δ neighborhood of the (left) half line restrictions of $\mathcal{R}(A)$. In other words, consider

$$\mathcal{R}_\delta^- = \{K_- : d_-(K_-, \mathcal{R}(A)_-) < \delta\}.$$

We then define $\Delta(K_-)$ for $K_- \in \mathcal{R}_\delta^-$ by setting $\Delta(K_-) = L_+$, where $L \in \mathcal{R}(A)$ is chosen such that $d_-(K_-, L_-) < \delta$; of course, there will be many such L for a given K_-, and then we just make an arbitrary choice. For $K_- \notin \mathcal{R}_\delta^-$, we can give $\Delta(K_-)$ any value whatsoever, or we can leave it undefined.

This function Δ has the required properties. To verify this, take x_0 so large that $d(x) < \delta$ for all $x \geq x_0$, with $d(x)$ defined as in Proposition 7.15.

So if now $x \geq x_0$ and $x \cdot H$ is a twisted shift of H of that length, then $d(x \cdot H, K) < \delta$ for some $K \in \omega(H)$. By Theorem 7.18, K will be in $\mathcal{R}(A)$. This shows that $(x \cdot H)_- \in \mathcal{R}_\delta^-$, and thus $\Delta((x \cdot H)_-) = L_+$ for some L as above. More explicitly, we have $L \in \mathcal{R}(A)$ also, and $d_-((x \cdot H)_-, L_-) < \delta$. Then $d_-(K_-, L_-) < 2\delta$, and now the choice of δ makes sure that $d_+(K_+, L_+) < \epsilon$, so $d_+((x \cdot H)_+, L_+) < 2\epsilon$, as required. □

The key ingredient to the proof of Theorem 7.18 will be the following result.

Theorem 7.21. *Consider any sequence of twisted shifts of $H \in \mathcal{C}(\mathbb{R})$ of lengths $x_n \to \infty$. Then for all Borel sets $A \subseteq \Sigma_{ac}^+(H)$, $|A| < \infty$, and $S \subseteq \mathbb{R}$ we have*

$$\lim_{n\to\infty} \left(\int_A \omega_{m_-(t;x_n \cdot H)}(-S)\, dt - \int_A \omega_{m_+(t;x_n \cdot H)}(S)\, dt \right) = 0.$$

Since I have stated this in a fairly general version, we need to address the issue of what happens here if m_+ or $m_- \equiv a \in \mathbb{R}_\infty$. In this case, we did not even define $\omega_{m(t)}(S)$. However, it is in fact quite easy to see that this is not a real problem: we can assume that $|\Sigma_{ac}^+| > 0$, or otherwise the theorem is not saying anything. But then H can not have a singular right half line (L, ∞), so m_+ is not constant, and m_- will not be either for all sufficiently large n.

I will prove Theorem 7.21 at the end of this section. Let us first see how this result gives us Theorem 7.18.

Proof of Theorem 7.18. Here, too, we may assume that H does not have a singular right half line, or else $|\Sigma_{ac}^+| = 0$, and the theorem is not claiming anything. Next, as observed, neither $\Sigma_{ac}^+(H)$ nor $\omega(H)$ will then depend on what H does on $(-\infty, 0)$, so we can modify H there and thus assume that $\Sigma_{ac}^- \supseteq \Sigma_{ac}^+$; for example, $H(x) = 1/2$ for $x < 0$ would work for this purpose. Then Im $m_\pm(t) > 0$ almost everywhere on $t \in \Sigma_{ac}^+$, and thus also $|R| < 1$ there.

Now let $K \in \omega(H)$, so $x_n \cdot H \to K$ for some sequence of twisted shifts of lengths $x_n \to \infty$. If at least one of $m_\pm(z; K)$ satisfies $m \equiv a \in \mathbb{R}_\infty$, then $K \equiv P_\alpha$, or else we would obtain $|R(t; K)| = 1$, but this contradicts Corollary 7.13.

So only the other case, when $m_\pm(z; K)$ are both genuine Herglotz functions, is nontrivial. We are then in case (a) of Theorem 7.5, so

$$m_\pm(z; x_n \cdot H) \to m_\pm(z; K)$$

in value distribution. Now combine this with Theorem 7.21. It follows that

$$\int_A \omega_{m_-(t;K)}(-S)\, dt = \int_A \omega_{m_+(t;K)}(S)\, dt$$

for all $A \subseteq \Sigma_{ac}^+$ of finite measure and all $S \subseteq \mathbb{R}$. As in Section 7.2, this implies that

$$\omega_{m_-(t;K)}(-S_j) = \omega_{m_+(t;K)}(S_j)$$

for almost every $t \in \Sigma_{ac}^+$ and for all intervals S_j with rational endpoints.

We may also assume here, at the expense of possibly having to remove another null set, that $m_\pm(t; K)$ both exist for these t and Im $m_\pm(t; K) > 0$. This latter requirement can be imposed because both m functions must have positive imaginary part almost everywhere on Σ_{ac}^+, or else it would again follow that $|R| = 1$.

In this situation, the mildly convoluted definition of $\omega_{m_\pm(t)}(S)$ simplifies, and we may simply compute this quantity as harmonic measure itself at the point $m_\pm \in \mathbb{C}^+$, as we observed earlier. It is then easy to show that a number $M \in \mathbb{C}^+$ is determined by the values of $\omega_M(S_j)$, $j \geq 1$, and $\omega_{x+iy}(-S) = \omega_{-x+iy}(S)$. Hence $-\overline{m_-(t; K)} = m_+(t; K)$ almost everywhere on Σ_{ac}^+, so $K \in \mathcal{R}(\Sigma_{ac}^+)$, by Proposition 7.8. □

Proof of Theorem 7.21. By compactness, it suffices to prove this for sequences of twisted shifts of lengths $x_n \to \infty$ for which $x_n \cdot H \to K$, for some $K \in \mathcal{C}(\mathbb{R})$. Then, exactly as in the proof of Theorem 7.18, it will again be convenient to modify H on $(-\infty, 0)$, if necessary, to make Im $m_- > 0$ almost everywhere on Σ_{ac}^+ also. This is admissible because if H_1 denotes such a modification, then $x_n \cdot H_1 \to K$ also, as we see from Lemma 7.17. Now Theorem 7.5 shows that what we are trying to prove is not affected by the modification.

A key tool in this proof will be the hyperbolic distance δ on $\mathbb{C}^+$. I already mentioned it briefly above; see (7.11). Here, it will be convenient to not work with δ itself, but rather with

$$\gamma(w,z) = \frac{|w-z|}{\sqrt{\operatorname{Im} w}\sqrt{\operatorname{Im} z}}.$$

This is not a metric because the triangle inequality fails, but it satisfies $\gamma^2 = \cosh 2\delta - 1$, so γ is a monotone function of δ, and $\gamma \to 0$ if and only if $\delta \to 0$, and this will be good enough for our purposes.

Holomorphic self-maps of $\mathbb{C}^+$ decrease hyperbolic distance, so if F is a Herglotz function, then $\gamma(F(w), F(z)) \le \gamma(w,z)$; if F is an automorphism, that is, F is a linear fractional transformation corresponding to the action of a $P \in SL(2,\mathbb{R})$, then we have equality here. Finally, γ controls the variation of harmonic measure in the sense that

$$|\omega_w(S) - \omega_z(S)| \le \gamma(w,z), \quad w,z \in \mathbb{C}^+. \tag{7.21}$$

To prove this, recall that $z \mapsto \omega_z(S)$ is a positive harmonic function on $z \in \mathbb{C}^+$ (or $\omega_z(S) \equiv 0$, if $|S| = 0$, but then of course everything is clear), so if $\alpha(z)$ denotes a harmonic conjugate, then $F(z) = \alpha(z) + i\omega(S)$ is a Herglotz function and thus

$$\gamma(w,z) \ge \gamma(F(w), F(z)) \ge \frac{|\omega_w(S) - \omega_z(S)|}{\sqrt{\omega_w(S)}\sqrt{\omega_z(S)}} \ge |\omega_w(S) - \omega_z(S)|,$$

since $\omega \le 1$.

After these preparations, we are now ready for the argument itself. Let $\epsilon > 0$ be given. We start by decomposing $A = A_0 \cup A_1 \cup \cdots \cup A_N$, and we choose these sets in such a way that $m_+(t) = m_+(t;H)$ is almost constant on each A_j, $j \ge 1$, and A_0 will serve as a wastebasket for those $t \in A$ at which certain undesired things happen. We first of all put all $t \in A$ at which $m_+(t)$ does not exist or does not have positive imaginary part into A_0; so far, $|A_0| = 0$. Next, pick (large enough) compact subsets $C \subseteq \mathbb{C}^+$, $C' \subseteq \mathbb{R}$, such that the set

$$A \setminus C' \cup \{t \in C' : m_+(t) \notin C\}$$

has measure $< \epsilon$. These points will also go into A_0, and this will then be our final choice of this set. Then decompose C into finitely many small subsets B_j of hyperbolic (that is, referring to γ) diameter less than ϵ, and put

$$A_j = \{t \in A \setminus A_0 : m_+(t) \in B_j\}.$$

We then also fix numbers $m_j \in B_j \subseteq \mathbb{C}^+$ for $j = 1, \ldots, N$. Then, by construction, $\gamma(m_+(t), m_j) < \epsilon$ for all $t \in A_j$.

It is then also true that $m_+(t; x\cdot H) = \lim_{y\to 0+} m_+(t+iy; x\cdot H)$ exists and lies in $\mathbb{C}^+$ for any twisted shift $x\cdot H$. This follows because the matrix function $S(z) = PT(x;z)$ that acts on $m_+(z;H)$ to implement the twisted shift is an entire function of z and $S(t) \in SL(2,\mathbb{R})$.

Furthermore, $m_+(t; x \cdot H) = S(t)m_+(t; H)$, so since $S(t) \in \mathrm{SL}(2, \mathbb{R})$ preserves γ, we then also have

$$\gamma(m_+(t; x \cdot H), PT(x; t)m_j) < \epsilon, \quad t \in A_j.$$

We can now use (7.21) and integrate over $t \in A_j$ to deduce that

$$\left| \int_{A_j} \omega_{m_+(t;x\cdot H)}(S)\, dt - \int_{A_j} \omega_{PT(x;t)m_j}(S)\, dt \right| < \epsilon |A_j| \tag{7.22}$$

for $j = 1, \ldots, N$.

We already observed earlier that changing S to $-S$ in $\omega_z(S)$ has the same effect as reflecting the point $z = x + iy$ about the imaginary axis and replacing it by $-x + iy = -\overline{z}$. Thus, since $PT(x; t)$ is real, we have

$$\omega_{PT(x;t)m_j}(S) = \omega_{-PT(x;t)\overline{m_j}}(-S).$$

We can write $-PT\overline{m_j} = IPTI(-\overline{m_j})$, with $I = \left(\begin{smallmatrix} 1 & 0 \\ 0 & -1 \end{smallmatrix}\right)$, as above. The matrix $IPT(x; z)I$ updates $m_-(z; H)$ under the twisted shift map $x \mapsto x \cdot H$. Of course, we are not applying it to m_- here, but to the constant $-\overline{m_j}$, but it will turn out that this does not really matter as $x \to \infty$: this evolution is focusing in the sense that any two initial values will eventually move almost in sync with each other. More precisely, this is true in the upper half plane $z \in \mathbb{C}^+$. I formulate this important statement as a separate result.

Lemma 7.22. *Let $H \in \mathcal{C}(\mathbb{R})$, $H \not\equiv P_\alpha$, on $(0, \infty)$. Consider a sequence of twisted shifts of H of lengths $x_n \to \infty$ and suppose that $x_n \cdot H \to K$. Denote the matrix functions that implement these maps by $S_n(z) = P_n T(x_n; z)$ and define $R_n(z) \in (0, \infty]$ for $z \in \mathbb{C}^+$ as the hyperbolic diameter of the disk $IS_n(z)I\overline{\mathbb{C}^+}$, that is,*

$$R_n(z) = \sup_{p,q \in \mathbb{C}^+} \gamma(IS_n(z)Ip, IS_n(z)Iq).$$

If K does not start with a singular half line $(-\infty, L)$, $L \geq 0$, then $R_n(z) \to 0$ locally uniformly on $z \in \mathbb{C}^+$.

Proof. If this is false, then there are $\delta > 0$ and a compact subset $C \subseteq \mathbb{C}^+$, such that

$$\gamma(IS_n(z_n)Ip_n, IS_n(z_n)Iq_n) \geq \delta \tag{7.23}$$

on a subsequence, which I have not made explicit in the notation, and for certain $z_n \in C$, $p_n, q_n \in \mathbb{C}^+$. For any $q \in \mathcal{F}$, the function $IS_n(z)Iq(z)$ is the m function m_- of the coefficient function that is the trace normed version of

$$H_n(t) = \begin{cases} P_n^{-1t} H(t + x_n) P_n^{-1}, & -x_n < t < 0, \\ H_q(t + x_n), & t < -x_n, \end{cases}$$

and here H_q of course denotes the $H \in \mathcal{C}_-$ that has q as its m function. By Lemma 7.17, $d(H_n, K_-) \to 0$ for any choice of $q = q_n(z)$. Thus $IS_nIq \to m_-(z;K)$ locally uniformly on $\mathbb{C}^+$, and in particular this holds for the two sequences of constant generalized Herglotz functions p_n, q_n from (7.23). On compact subsets of $\mathbb{C}^+$, away from the boundary, the hyperbolic and Euclidean distances are comparable, so the only way to avoid a contradiction would be to have $m_-(z;K) \equiv a \in \mathbb{R}_\infty$. However, this we explicitly ruled out, so (7.23) is not tenable. □

To apply the lemma to the present analysis, we must move from $t \in \mathbb{R}$ into the upper half plane, and to this end, we invoke Theorem 7.6. Pick a $y > 0$ such that

$$\left| \int_{A_j} \omega_{F(t+iy)}(S)\,dt - \int_{A_j} \omega_{F(t)}(S)\,dt \right| < \epsilon |A_j| \tag{7.24}$$

for $j = 1, \dots, N$ and all $F \in \mathcal{F}$, $S \subseteq \mathbb{R}$. Recall that we earlier fixed a sequence $x_n \to \infty$ such that $x_n \cdot H \to K$, as required in Lemma 7.22. If this K has a singular half line and $|\Sigma_{ac}^+(H)| > 0$, then $K \equiv P_\alpha$ on $\mathbb{R}$, by Theorem 7.12, and then the claim of Theorem 7.21 along this subsequence follows quickly from Theorem 7.5(b), (c).

In the other case, when $K \not\equiv P_\alpha$, we obtain

$$\gamma(IP_nT(x_n; t+iy)I(-\overline{m_j}), m_-(t+iy; x_n \cdot H)) < \epsilon$$

for all sufficiently large n, uniformly in $t \in \bigcup_{1\le j\le N} A_j$. By (7.21), this implies that

$$\left| \int_{A_j} \omega_{IP_nT(x_n;t+iy)I(-\overline{m_j})}(-S)\,dt - \int_{A_j} \omega_{m_-(t+iy;x_n\cdot H)}(-S)\,dt \right| < \epsilon |A_j|.$$

Now combine this with (7.22), (7.24) and recall that $|A_0| < \epsilon$. We conclude that

$$\left| \int_{A} \omega_{m_+(t;x_n\cdot H)}(S)\,dt - \int_{A} \omega_{m_-(t;x_n\cdot H)}(-S)\,dt \right| < 2\epsilon + 4\epsilon |A|$$

for all large n. □

7.6 Notes

This chapter is based mainly on my own recent work [55–57]. I take the opportunity to discuss various issues here, mostly related to canonical systems with singular half lines or even consisting of just a single singular interval, that were not given proper attention in the original works.

Section 7.1. Theorem 7.2 (and its companion result Theorem 7.7) is from [57], but was given an unnecessarily complicated proof there because I failed to observe the neat transformation formula from Theorem 7.1. Like here, it was formulated there as a statement about the multiplication operators M_t in the $L^2(\mathbb{R}, \rho)$ spaces of the spectral representation. It is also natural to wonder if the relations themselves (not just their operator parts) are unitarily equivalent. Of course, asking about unitary equivalence of multi-valued parts $(0, \mathcal{S}(0))$ is just asking about the dimension of $\mathcal{S}(0)$, and for a canonical system, $\mathcal{S}(0)$ is the zero space or infinite-dimensional. So we are asking if a non-zero multi-valued part of a canonical system will also give a non-zero multi-valued part to all canonical systems that can be reached from the first one by a Toda map. Or, equivalently, we can ask if a singular interval somewhere will also lead to at least one singular interval of all canonical systems reachable by a Toda map. I do not know if this is true.

Section 7.2. Theorem 7.4 is from [56] (though only genuine Herglotz functions were considered in that reference). Theorem 7.5 as stated is taken from [55], but the result is really contained (though not formulated explicitly there) in the work of Pearson [9, 10, 50, 51], who first got interested in the quantities $\int_A \omega_{F(t)}(S)\, dt$ in the context of his work on spectral averaging. The name *value distribution* is quite descriptive here, but it has come under criticism since it is also used for Nevanlinna's completely unrelated theory of meromorphic functions.

The stunning Theorem 7.6 is due to Breimesser and Pearson [10].

Section 7.4. Theorem 7.12 is from [56]. The semi-continuity of Σ_{ac}, in a more specialized setting for shifts of Schrödinger operators or Jacobi matrices, was first observed by Last and Simon [45].

Section 7.5. Theorem 7.18 was first proved for Jacobi matrices in [55], with a very similar proof. A version for canonical systems very similar to the one proved here is given in [56], and if only shifts are considered, then the result may be found in [2]. The presentation I give here improves on [56] in that the degenerate canonical systems $H \equiv P_\alpha$ are now given proper attention. The key ingredient is always a result of the type of Theorem 7.21, which, for Schrödinger operators, is due to Breimesser and Pearson [9]. Again, the proof does not really depend much on which type of equation is being considered, except perhaps for the part that I formulated as Lemma 7.22 here. Instead of this, [9] uses a more concrete (and quite ingenious) calculation that does work with the equation itself; in the case of Jacobi matrices, this concrete argument becomes rather simple (see [55]).

Given a result like Theorem 7.18, the ideas on how to deduce the predictability of coefficient functions with some absolutely continuous spectrum that are expressed in Theorem 7.20 are due to Kotani [39]. In fact, Kotani's well-known theory of ergodic Schrödinger operators [37–40] provided an important source of inspiration for much of the work reported on in this section. With some extra work, one can show that the oracle can be chosen as a continuous function; see [54] for a discussion of this issue in a slightly different setting.

Bibliography

[1] K. Acharya, An alternate proof of the de Branges theorem on canonical systems, *ISRN Math. Anal.* (2014), 7 pp.

[2] K. Acharya, Remling's theorem on canonical systems, *J. Math. Phys.* **57** (2016), 11 pp.

[3] D. Alpay (editor), Operator Theory, Springer, Basel, 2015.

[4] D. Z. Arov and H. Dym, *J*-inner matrix functions, interpolation and inverse problems for canonical systems, I. Foundations, *Integral Equ. Oper. Theory* **29** (1997), 373–454.

[5] D. Z. Arov and H. Dym, Some remarks on the inverse monodromy problem for 2×2 canonical differential systems, in Operator Theory and Analysis, Oper. Theory, Adv. Appl., Vol. 122, Birkhäuser, Basel, 2001, 53–87.

[6] A. Baranov and H. Woracek, Majorization in de Branges spaces, I. Representability of subspaces, *J. Funct. Anal.* **258** (2010), 2601–2636.

[7] J. Behrndt, S. Hassi, H. de Snoo, and R. Wietsma, Square-integrable solutions and Weyl functions for singular canonical systems, *Math. Nachr.* **284** (2011), 1334–1384.

[8] R. Boas, Entire Functions, Academic Press, New York, 1954.

[9] S. Breimesser and D. Pearson, Asymptotic value distribution for solutions of the Schrödinger equation, *Math. Phys. Anal. Geom.* **3** (2000), 385–403.

[10] S. Breimesser and D. Pearson, Geometrical aspects of spectral theory and value distribution for Herglotz functions, *Math. Phys. Anal. Geom.* **6** (2003), 29–57.

[11] T. Carleman, Sur une inegalite differentielle dans la theorie des fonctions analytiques, *C. R. Acad. Sci. Paris* **196** (1933) 995–997.

[12] E. A. Coddington and N. Levinson, Theory of Ordinary Differential Equations, McGraw-Hill, New York, 1955.

[13] L. de Branges, Some Hilbert spaces of entire functions, *Trans. Am. Math. Soc.* **96** (1960), 259–295.

[14] L. de Branges, Some Hilbert spaces of entire functions II, *Trans. Am. Math. Soc.* **99** (1961), 118–152.

[15] L. de Branges, Some Hilbert spaces of entire functions III, *Trans. Am. Math. Soc.* **100** (1961), 73–115.

[16] L. de Branges, Some Hilbert spaces of entire functions IV, *Trans. Am. Math. Soc.* **105** (1961), 43–83.

[17] L. de Branges, Hilbert Spaces of Entire Functions, Prentice-Hall, Englewood Cliffs, 1968.

[18] H. de Snoo and H. Winkler, Canonical systems of differential equations with self-adjoint interface conditions on graphs, *Proc. R. Soc. Edinb. A* **135** (2005), 297–315.

[19] H. de Snoo and H. Winkler, Two-dimensional trace-normed canonical systems of differential equations and self-adjoint interface conditions, *Integral Equ. Oper. Theory* **51** (2005), 73–108.

[20] V. A. Derkach and M. M. Malamud, The extension theory of hermitian operators and the moment problem, *J. Math. Sci.* **73** (1995), 141–242.

[21] A. Dijksma and H. de Snoo, Self-adjoint extensions of symmetric subspaces, *Pac. J. Math.* **54** (1974), 71–100.

[22] P. Duren, Theory of H^p Spaces, Dover reprint, Mineola, 2000.

[23] H. Dym and H. P. McKean, Gaussian Processes, Function Theory, and the Inverse Spectral Problem, Dover reprint, Mineola, 2008.

[24] H. Dym and S. Sarkar, Multiplication operators with deficiency indices (p, p) and sampling formulas in reproducing kernel Hilbert spaces of entire vector valued functions, *J. Funct. Anal.* **273** (2017), 3671–3718.

[25] G. Folland, Real Analysis, Wiley, New York, 1999.

https://doi.org/10.1515/9783110563238-008

[26] J. Garnett, Bounded Analytic Functions, Grad. Texts Math., Vol. 235, Springer, New York, 2007.
[27] I. C. Gohberg and M. G. Krein, Theory and Applications of Volterra Operators in Hilbert Space, Transl. Math. Monogr., Vol. 24, American Mathematical Society, Providence, 1970.
[28] M. L. Gorbachuk and V. I. Gorbachuk, M. G. Krein's Lectures on Entire Operators, Oper. Theory, Adv. Appl., Vol. 97, Birkhäuser Verlag, Basel (1997).
[29] S. Hassi, H. de Snoo, and H. Winkler, Boundary value problems for two-dimensional canonical systems, *Integral Equ. Oper. Theory* **36** (2000), 445–479.
[30] M. Heins, Selected Topics in the Classical Theory of Functions of a Complex Variable, Holt, Reinhart and Winston, New York, 1962.
[31] I. S. Kac, On the multiplicity of the spectrum of a second order differential operator, *Sov. Math.* **3** (1962), 1035–1039.
[32] I. S. Kac, The Hamburger moment problem as part of spectral theory of canonical systems, *Funct. Anal. Appl.* **33** (1999), 228–230.
[33] I. S. Kac, Linear relations generated by a canonical differential equation of dimension 2 and eigenfunction expansions, *St. Petersburg Math. J.* **14** (2003), 429–452.
[34] M. Kaltenbäck, H. Winkler, and H. Woracek, Strings, dual strings, and related canonical systems, *Math. Nachr.* **280** (2007), 1518–1536.
[35] P. Koosis, Introduction to H^p spaces, London Math. Soc. Lect. Notes, Vol. 40, Cambridge University Press, Cambridge, 1980.
[36] P. Koosis, The Logarithmic Integral I, Cambridge University Press, Cambridge, 1988.
[37] S. Kotani, Ljapunov indices determine absolutely continuous spectra of stationary random one-dimensional Schrödinger operators, in Stochastic Analysis (Katata/Kyoto 1982), N.-Holl. Math. Libr., Vol. 32, North-Holland, Amsterdam, 1984, 225–247.
[38] S. Kotani, One-dimensional random Schrödinger operators and Herglotz functions, in Probabilistic Methods in Mathematical Physics (Katata/Kyoto 1985), Academic Press, Boston, 1987, 219–250.
[39] S. Kotani, Jacobi matrices with random potentials taking finitely many values, *Rev. Math. Phys.* **1** (1989), 129–133.
[40] S. Kotani, Generalized Floquet theory for stationary Schrödinger operators in one dimension, *Chaos Solitons Fractals* **8** (1997), 1817–1854.
[41] M. G. Krein, Determination of the density of a nonhomogeneous symmetric cord by its frequency spectrum (Russian), *Dokl. Akad. Nauk SSSR* **76** (1951), 345–348.
[42] M. G. Krein and H. Langer, Continuation of Hermitian positive definite functions and related questions, *Integral Equ. Oper. Theory* **78** (2014), 1–69.
[43] H. Langer and B. Textorius, *L*-resolvent matrices of symmetric linear relations with equal defect numbers; applications to canonical differential relations, *Integral Equ. Oper. Theory* **5** (1982), 208–243.
[44] M. Langer and H. Woracek, A local inverse spectral theorem for Hamiltonian systems, *Inverse Probl.* **27** (2011), 17 pp.
[45] Y. Last and B. Simon, Eigenfunctions, transfer matrices, and absolutely continuous spectrum of one-dimensional Schrödinger operators, *Invent. Math.* **135** (1999), 329–367.
[46] B. Levin, Distributions of Zeros of Entire Functions, Transl. Math. Monogr., Vol. 5, American Mathematical Society, Providence, 1964.
[47] B. Levin, Lectures on Entire Functions, Transl. Math. Monogr., Vol. 150, American Mathematical Society, Providence, 1996.
[48] D. Linghu, Chains of non-regular de Branges spaces, PhD thesis, Caltech, 2015.
[49] R. Nevanlinna, Aymptotische Entwicklungen beschränkter Funktionen und das Stieltjessche Momentenproblem, *Ann. Acad. Sci. Fenn. A* **18** (1922).

[50] D. Pearson, Value distribution and spectral analysis of differential operators, *J. Phys. A* **26** (1993), 4067–4080.
[51] D. Pearson, Value distribution and spectral theory, *Proc. Lond. Math. Soc.* **68** (1994), 127–144.
[52] V. P. Potapov, On the multiplicative structure of *J*-contractive matrix functions, *Am. Math. Soc. Transl. (2)* **15** (1960), 131–243.
[53] C. Remling, Schrödinger operators and de Branges spaces, *J. Funct. Anal.* **196** (2002), 323–394.
[54] C. Remling, The absolutely continuous spectrum of one-dimensional Schrödinger operators, *Math. Phys. Anal. Geom.* **10** (2007), 359–373.
[55] C. Remling, The absolutely continuous spectrum of Jacobi matrices, *Ann. Math.* **174** (2011), 125–171.
[56] C. Remling, Generalized reflection coefficients, *Commun. Math. Phys.* **337** (2015), 1011–1026.
[57] C. Remling, Toda maps, cocycles, and canonical systems, *J. Spectr. Theory*, in press.
[58] R. Romanov, Jacobi matrices and de Branges spaces, in D. Alpay (editor), Operator Theory, Springer, Basel, 2015, 609–621.
[59] R. Romanov, Canonical systems and de Branges spaces, London Mathematical Society Lecture Notes, in press.
[60] M. Rosenblum and J. Rovnyak, Topics in Hardy Classes and Univalent Functions, Birkhäuser Verlag, Basel, 1994.
[61] W. Rudin, Functional Analysis, McGraw-Hill, Singapore, 1991.
[62] L. A. Sakhnovich, Spectral Theory of Canonical Differential Systems: Method of Operator Identities, Oper. Theory, Adv. Appl., Vol. 107, Birkhäuser Verlag, Basel, 1999.
[63] S. L. Segal, Nine Introductions in Complex Analysis, North-Holland Mathematical Studies 208, Elsevier, Amsterdam, 2008.
[64] B. Simon, The classical moment problem as a self-adjoint finite difference operator, *Adv. Math.* **137** (1998), 82–203.
[65] J. Weidmann, Spectral Theory of Ordinary Differential Operators, Springer Lect. Notes, Vol. 1258, Springer, Berlin, 1987.
[66] H. Weyl, Über gewöhnliche Differentialgleichungen mit Singularitäten und die zugehörigen Entwicklungen willkürlicher Funktionen, *Math. Ann.* **68** (1910), 220–269.
[67] H. Winkler, The inverse spectral problem for canonical systems, *Integral Equ. Oper. Theory* **22** (1995), 360–374.
[68] H. Winkler, Canonical systems with a semibounded spectrum, *Oper. Theory, Adv. Appl.* **106** (1998), 397–417.
[69] H. Winkler, Spectral estimations for canonical systems, *Math. Nachr.* **220** (2000), 115–141.
[70] H. Winkler and H. Woracek, On semibounded canonical systems, *Linear Algebra Appl.* **429** (2008), 1082–1092.
[71] H. Woracek, De Branges spaces and growth aspects, in D. Alpay (editor), Operator Theory, Springer, Basel, 2015, 489–523.

Index

De Gruyter Studies in Mathematics

Volume 69
Derek K. Thomas, Nikola Tuneski, Allu Vasudevarao
Univalent Functions. A Primer
ISBN 978-3-11-056009-1, e-ISBN 978-3-11-056096-1, e-ISBN (ePUB) 978-3-11-056012-1

Volume 68/1
Timofey V. Rodionov, Valeriy K. Zakharov
Set, Functions, Measures. Volume 1: Fundamentals of Set and Number Theory
ISBN 978-3-11-055008-5, e-ISBN 978-3-11-055094-8

Volume 68/2
Alexander V. Mikhalev, Timofey V. Rodionov, Valeriy K. Zakharov
Set, Functions, Measures. Volume 2: Fundamentals of Functions and Measure Theory
ISBN 978-3-11-055009-2, e-ISBN 978-3-11-055096-2

Volume 67
Alexei Kulik
Ergodic Behavior of Markov Processes. With Applications to Limit Theorems
ISBN 978-3-11-045870-1, e-ISBN 978-3-11-045893-0

Volume 66
Igor V. Nikolaev
Noncommutative Geometry. A Functorial Approach, 2017
ISBN 978-3-11-054317-9, e-ISBN 978-3-11-054525-8

Volume 65
Günter Mayer
Interval Analysis. And Automatic Result Verification, 2017
ISBN 978-3-11-050063-9, e-ISBN 978-3-11-049946-9

Volume 64
Dorina Mitrea, Irina Mitrea, Marius Mitrea, Michael Taylor
The Hodge-Laplacian. Boundary Value Problems on Riemannian Manifolds, 2016
ISBN 978-3-11-048266-9, e-ISBN 978-3-11-048438-0

Volume 63
Evguenii A. Rakhmanov
Orthogonal Polynomials, 2016
ISBN 978-3-11-031385-7, e-ISBN 978-3-11-031386-4

www.degruyter.com

Bei Fragen zur Produktsicherheit wenden Sie sich bitte an:
If you have any questions regarding product safety,
please contact:

Walter de Gruyter GmbH
Genthiner Straße 13
10785 Berlin
productsafety@degruyterbrill.com